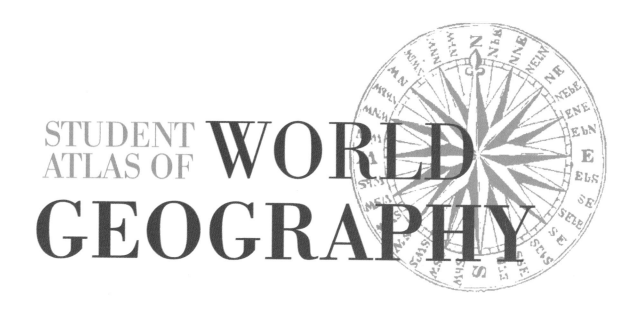

STUDENT ATLAS OF WORLD GEOGRAPHY

John L. Allen
University of Connecticut

Dushkin/McGraw-Hill

A Division of The **McGraw·Hill** Companies

Book Team

Vice President & Publisher   *Jeffrey L. Hahn*
Production Manager   *Brenda S. Filley*
Designer   *Charles Vitelli*
Typesetting Supervisor   *Juliana Arbo*
Proofreader   *Diane Barker*
Cover design   *Tom Goddard*
Cartography   *Carto-Graphics, Eau Claire, WI*

We would like to thank Digital Wisdom Incorporated for allowing us to use their Mountain High Maps cartography software. This software was used to create maps 62, 63, 65, 66, 68, 69, 71, 72, 74, 75, 77, 78.

Dushkin/McGraw-Hill

*A Division of The **McGraw·Hill** Companies*

Library of Congress Catalog Card Number: 00-00000

ISBN 0-07-228568-0

10 9 8 7 6 5 4 3 2

To the Student

The study of geography has become an increasingly important part of the curriculum in secondary schools and institutions of higher education over the last decade. This trend, a most welcome one from the standpoint of geographers, has begun to address the massive problem of "geographic illiteracy" that has characterized the United States, almost alone among the world's developed nations. When a number of international comparative studies on world geography were undertaken, beginning in the 1970s, it became apparent that most American students fell far short of their counterparts in Europe, Russia, Canada, Australia, and Japan in their abilities to recognize geographic location, to identify countries or regions on maps, or to explain the significance of such key geographic phenomena such as population distribution, economic or urban location, or the availability of natural resources. Indeed, many American students could not even locate the United States on world maps, let alone countries like France, or Indonesia, or Nigeria. This atlas, and the texts it is intended to accompany, is a small part of the process of attempting to increase the geographic literacy of American students. As the true meaning of "the global community" becomes more apparent, such an increase in geographic awareness is not only important but necessary.

The maps in the *Student Atlas of World Geography* are designed to introduce you to the patterns or "spatial distribution" of the wide variety of human and physical features of the earth's surface and to help you understand the relationships between these patterns. We call such relationships "spatial correlation" and whenever you compare the patterns made by two or more phenomena that exist at or near the earth's surface–the distribution of human population and the types of climate, for example–you are engaging in spatial correlation. The maps are not perfect representations of reality–no maps ever are–but they do represent "models" or approximations of the real world that should aid in your understanding of the patterns they portray. A model is a simplification of reality and it is often shown at a size that is a specific proportion of the reality it is intended to represent. An HO gauge model railroad boxcar, for example, is 1/87th the size of the real boxcar it represents. Similarly, maps are proportional representations of earth space. The numbers at the bottom of the maps in your atlas indicate the size of the proportion–either 1/125,000,000 or 1/180,000,000. These numbers tell us the "scale" or proportional representation of reality shown by the map. And just as the boxcar at a 1/87 scale cannot be as complicated as the real thing, neither can maps be as complicated as the real world. Maps are generalizations, then, of the geographic features they illustrate and the smaller the scale, the greater the level of generalization. The maps are all drawn using a type of process called a "projection" whereby the curved or spherical surface of the earth is converted to a flat piece of paper. Because such a conversion is impossible without some distortion of the size or shape of features on the surface of the earth (a globe is the only type of map that does not produce distortion), different types of projections are designed for different purposes. The projection used for this atlas is a modified equal-area projection. This means that the sizes of areas on the maps are very close to "true," that is, on a map of a 1 to 125,000,000 scale, the size of any area will be about 1/125,000,000th the size of that area in the real world. The equal-area projection makes comparison of areal phenomena not only easier but quantitatively more valid.

Like the maps, the data sets in the atlas are intended to enable you to make comparisons between the distributions of different geographic features (for example, population growth and literacy rates) and to understand the character of the geographic variation in a single geographic feature. In many instances, the data in the tables of your atlas are the same data that have been used to produce the maps. At the very outset of your study of this atlas, you should be aware of some limitations of the data tables. In some instances, there may be data missing from a table. In such cases, the cause may represent the failure of a country to report information to a central international body (like the United Nations or the World Bank), or it may mean that the shifting of political boundaries and changed responsibility for reporting data have caused some countries (for example, those countries that made up the former Soviet Union or the former Yugoslavia) to delay their reports. It is always our aim to use the most up-to-date data that is possible. Subsequent editions of this atlas will have increased data on countries like Slovenia, Ukraine, or Uzbekistan when it becomes available. In the meantime, as events continue to restructure our world, it's an exciting time to be a student of world geography!

You will find your study of this atlas more productive if you study the maps and tables on the following pages in the context of the five distinct themes that have been developed as part of the increasing awareness of the importance of geographic education:

1. *Location: Where Is It?* This theme offers a starting point from which you discover the precise location of places in both absolute terms (the latitude and longitude of a place) and in relative terms (the location of a place in relation to the location of other places). When you think of location, you should automatically think of both forms. Knowing something about absolute location will help you to understand a variety of features of physical geography, since such key elements are so closely related to their position on the earth. But it is equally important to think of location in relative terms. The location of places in relation to other places is often more important as a determinant of social, economic, and cultural characteristics than the factors of physical geography.

2. *Place: What Is It Like?* This theme investigates the political, economic, cultural, environmental, and other characteristics that give a place its identity. You should seek to understand the similarities and differences of places by exploring their basic characteristics. Why are some places with similar environmental characteristics so very different in economic, cultural, social, and political ways? Why are other places with such different environmental characteristics so seemingly alike in terms of their institutions, their economies, and their cultures?

3. *Human/Environment Interactions: How Is the Landscape Shaped?* This theme illustrates the ways in which people respond to and modify their environments. Certainly the environment is an important factor in influencing human activities and behavior. But the characteristics of the environment do not exert a controlling influence over human activities; they only provide a set of alternatives from which different cultures, in different times, make their choices. Observe the relationship between the basic elements of physical geography such as climate and terrain and the host of ways in which humans have used the land surfaces of the world.

4. *Movement: How Do People Stay in Touch?* This theme examines the transportation and communication systems that link people and places. Movement or "spatial interaction" is the chief mechanism for the spread of ideas and innovations from one place to another. It is spatial interaction that

validates the old cliché, "the world is getting smaller." We find McDonald's restaurants in Tokyo and Honda automobiles in New York City because of spatial interaction. Advanced transportation and communication systems have transformed the world into which your parents were born. And the world your children will be born into will be very different from your world. None of this would happen without the force of movement or spatial interaction.

5. *Regions: Worlds Within a World.* This theme helps to organize knowledge about the land and its people. The world consists of a mosaic of "regions" or areas that are somehow different and distinctive from other areas. The region of Anglo-America (the United States and Canada) is, for example, different enough from the region of Western Europe that geographers clearly identify them as two unique and separate areas. Yet despite their differences, Anglo-Americans and Europeans share a number of similarities: common cultural backgrounds, comparable economic patterns, shared religious traditions, and even some shared physical environmental characteristics. Conversely, although the regions of Anglo-America and Eastern Asia are also easily distinguished as distinctive units of the Earth's surface, they have a greater number of shared physical environmental characteristics. But those who live in Anglo-America and Eastern Asia have fewer similarities and more differences between them than is the case with Anglo-America and Western Europe: different cultural traditions, different institutions, different linguistic and religious patterns. An understanding of both the differences and similarities between regions like Anglo-America and Europe on the one hand, or Anglo-America and Eastern Asia on the other, will help you to understand much that has happened in the human past or that is currently transpiring in the world around you. At the very least, an understanding of regional similarities and differences will help you to interpret what you read on the front page of your daily newspaper or view on the evening news report on your television set.

Not all of these themes will be immediately apparent on each of the maps and tables in this atlas. But if you study the contents of *Student Atlas of World Geography,* along with the reading of your text and think about the five themes, maps and tables and text will complement one another and improve your understanding of global geography.

John L. Allen

About the Author

John L. Allen is Professor of Geography at the University of Connecticut where he has taught since 1967. He is a native of Wyoming and received his bachelor's degree in 1963 and his M.A. in 1964 from the University of Wyoming, and in 1969 he received his Ph.D. from Clark University. His special areas of interest are perceptions of the environment and the impact of human societies on environmental systems. Dr. Allen is the author and editor of many books and articles as well as several other student atlases, including the best selling *Student Atlas of World Politics*.

Acknowledgments

The author wishes to recognize with gratitude the advice, suggestions, and general assistance of the following reviewers:

Robert Bednarz
Texas A & M University

Paul B. Frederic
University of Maine at Farmington

James F. Fryman
University of Northern Iowa

Lloyd E. Hudman
Brigham Young University

Artimus Keiffer
Indiana University-Purdue University at Indianapolis

Richard L. Krol
Kean College of New Jersey

Robert Larson
Indiana State University

Elizabeth J. Leppman
St. Cloud State University

Mark Lowry II
United States Military Academy at West Point

Max Lu
Kansas State University

Taylor E. Mack
Mississippi State University

Patrick McGreevy
Clarion University

Tyrel G. Moore
University of North Carolina at Charlotte

David J. Nemeth
The University of Toledo

Emmett Panzella
Point Park College

Jefferson S. Rogers
University of Tennessee at Martin

Barbara J. Rusnak
United States Air Force Academy

Mark Simpson
University of Tennessee at Martin

Table of Contents

Part VII. World Regions   85

North America

South America

Europe

Africa

Asia

Australasia

Part I

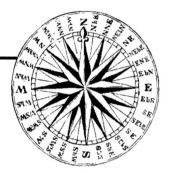

Global Physical Patterns

Map ⬚1 World Countries

The international system includes the political units called "states" or countries as the most important component. The boundaries of countries are the primary source of political division in the world and for most people nationalism is the strongest source of political identity. State boundaries are an important indicator of cultural, linguistic, economic, and other geographic divisions as well, and the states themselves normally serve as the base level for which most global statistics are available. The subfield of geography known as "Political Geography" has as its primary concern the geographic or spatial character of this international system and its components.

Scale: 1 to 125,000,000

Note: All world maps are Robinson projection.

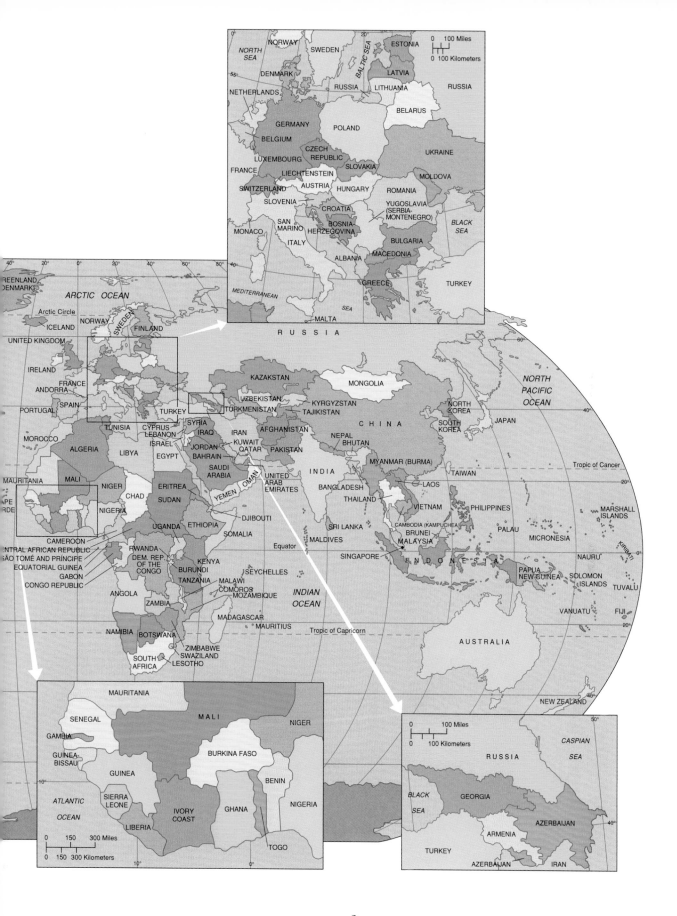

Map 2a Average Annual Precipitation

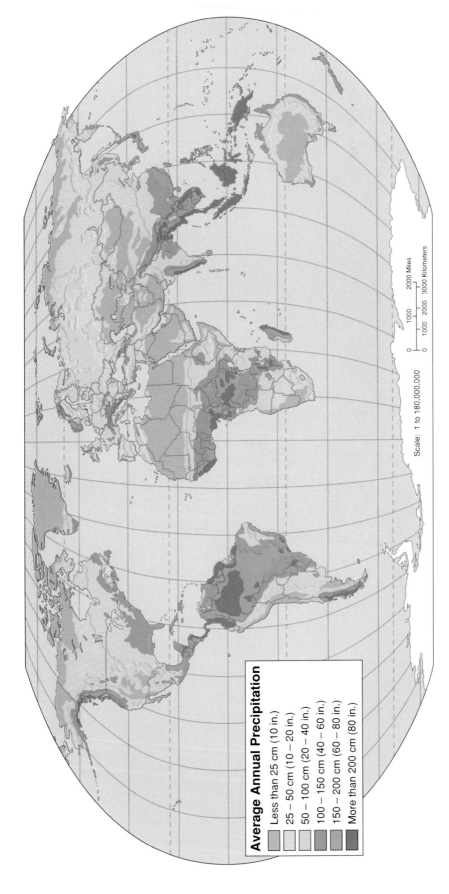

Average Annual Precipitation

- Less than 25 cm (10 in.)
- 25 − 50 cm (10 − 20 in.)
- 50 − 100 cm (20 − 40 in.)
- 100 − 150 cm (40 − 60 in.)
- 150 − 200 cm (60 − 80 in.)
- More than 200 cm (80 in.)

Scale: 1 to 180,000,000

0   1000   2000 Miles

0 1000 2000 3000 Kilometers

The two most important physical geographic variables are precipitation and temperature, the essential elements of weather and climate. Precipitation is a conditioner of both soil type and vegetation. More than any other single environmental element, it influences where people do or do not live. Water is the most precious resource available to humans, and water availability is largely a function of precipitation. Water availability is also a function of several precipitation variables that do not appear on this map: the seasonal distribution of precipitation (is precipitation or drought concentrated in a particular season?), the ratio between precipitation and temperature (how much of the water that comes to the earth in the form of precipitation is lost through mechanisms such as evaporation and transpiration that are a function of temperature?), and the annual variability of precipitation (how much do annual precipitation totals for a place or region tend to vary from the "normal" or average precipitation?). In order to obtain a complete understanding of precipitation, these variables should be examined along with the more general data presented on this map. The study of precipitation and other climatic elements is the concern of the branch of physical geography called "climatology."

Map 2b Seasonal Average Precipitation–November through April

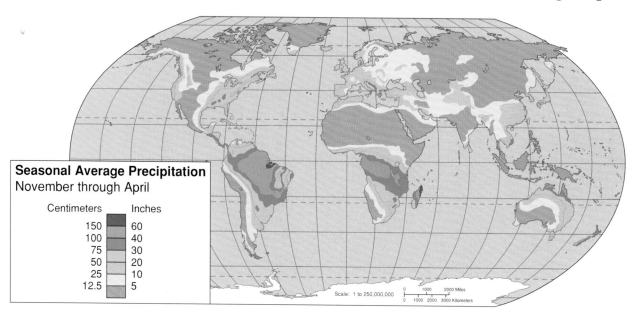

Seasonal Average Precipitation
November through April

Centimeters	Inches
150	60
100	40
75	30
50	20
25	10
12.5	5

Scale: 1 to 250,000,000

Map 2c Seasonal Average Precipitation–May through October

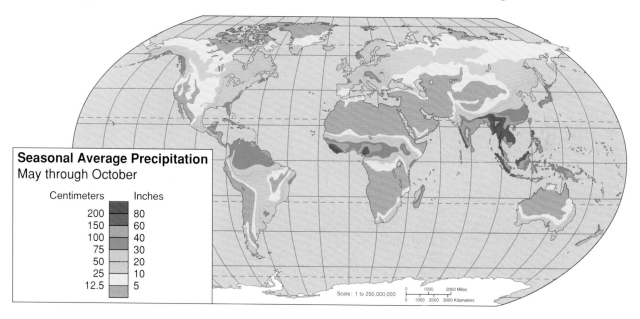

Seasonal Average Precipitation
May through October

Centimeters	Inches
200	80
150	60
100	40
75	30
50	20
25	10
12.5	5

Scale: 1 to 250,000,000

Seasonal average precipitation is nearly as important as annual precipitation totals in determining the habitability of an area. Critical factors are such things as whether precipitation coincides with the growing season and thus facilitates agriculture or during the winter when it is less effective in aiding plant growth, and whether precipitation occurs during summer with its higher water loss through evaporation and transpiration or during the winter when more of it can go into storage. Several of the world's great climate zones have pronounced seasonal precipitation rhythms. The tropical and subtropical savanna grasslands have a long winter dry season and abundant precipitation in the summer. The Mediterranean climate is the only major climate with a marked dry season during the summer, making agriculture possible only through irrigation or other adjustments to cope with drought during the period of plant growth. And the great monsoon climates of south and southeast Asia have their winter dry season and summer rain that has conditioned the development of Asian agriculture and the rhythms of Asian life.

Map 3a Temperature Regions and Ocean Currents

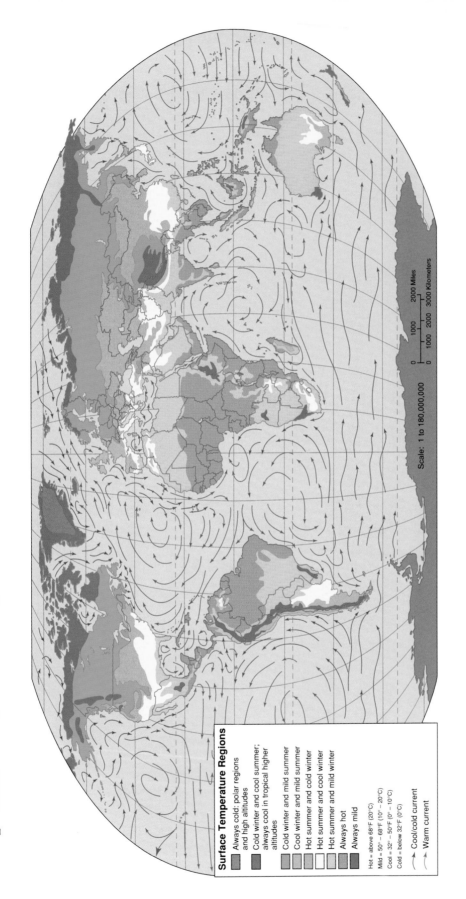

Surface Temperature Regions

- Always cold: polar regions and high altitudes
- Cold winter and cool summer; always cool in tropical higher altitudes
- Cold winter and mild summer
- Cool winter and mild summer
- Hot summer and cold winter
- Hot summer and cool winter
- Hot summer and mild winter
- Always hot
- Always mild

Hot = above 68°F (20°C)
Mild = 50° – 68°F (10° – 20°C)
Cool = 32° – 50°F (0° – 10°C)
Cold = below 32°F (0°C)

↗ Cool/cold current
↗ Warm current

Scale: 1 to 180,000,000

0   1000   2000 Miles
0   1000   2000   3000 Kilometers

Along with precipitation, temperature is one of the two most important environmental variables, defining the climatic conditions so essential for the distribution of such human activities as agriculture and the distribution of the human population. The seasonal rhythm of temperature, including such measures as the average annual temperature range (difference between the average temperature of the warmest month and that of the coldest month), is an additional variable not shown on the map but, like the seasonality of precipitation, should be a part of any comprehensive study of climate. The ocean currents illustrated exert a significant influence over the climate of adjacent regions and are the most important mechanism for redistributing surplus heat from the equatorial region into middle and high latitudes. Physical geographers known as "climatologists" study the phenomenon of temperature and related climatic characteristics.

Map 3b Average July Temperature

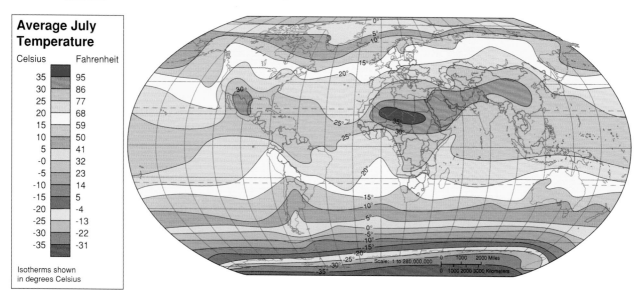

Average July Temperature

Celsius	Fahrenheit
35	95
30	86
25	77
20	68
15	59
10	50
5	41
-0	32
-5	23
-10	14
-15	5
-20	-4
-25	-13
-30	-22
-35	-31

Isotherms shown in degrees Celsius

Map 3c Average January Temperature

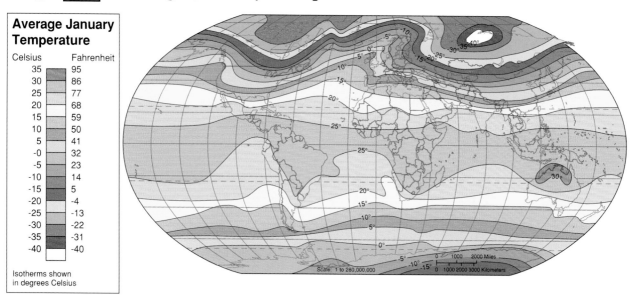

Average January Temperature

Celsius	Fahrenheit
35	95
30	86
25	77
20	68
15	59
10	50
5	41
-0	32
-5	23
-10	14
-15	5
-20	-4
-25	-13
-30	-22
-35	-31
-40	-40

Isotherms shown in degrees Celsius

Where moisture availability tends to mark the seasons in the tropics and subtropics, in the mid-latitudes, seasons are defined by temperature. Temperature is determined by latitudinal position, by altitude or elevation above sea level, and by the location of a place relative to the world's land masses and oceans. The most important of these controls is latitude, and temperatures generally become lower with increasing latitude. However, proximity to water tends to moderate temperature extremes, and "maritime" climates influenced by the oceans will be warmer in the winter and cooler in the summer than continental climates in the same general latitude. Maritime climates will also show smaller temperature ranges,

the difference between January and July temperatures, while climates of continental interiors, far from the moderating influences of the oceans, will tend to have great temperature ranges. In the northern hemisphere, where there are both large land masses and oceans, the range is great. But in the southern hemisphere, dominated by water and, hence, by the more moderate maritime air masses, the temperature range is comparatively small. Significant temperature departures from the "normal" produced by latitude may also be the result of elevation. With exceptions, lower temperatures produced by topography are difficult to see on maps of this scale.

Map 4a Atmospheric Pressure and Predominant Surface Winds–January

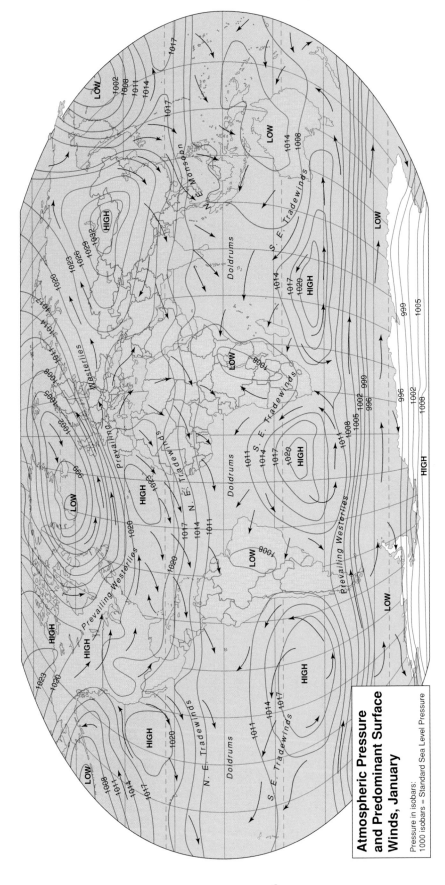

Atmospheric Pressure and Predominant Surface Winds, January

Pressure in isobars:
1000 isobars = Standard Sea Level Pressure

– 8 –

Map 4b Atmospheric Pressure and Predominant Surface Winds–July

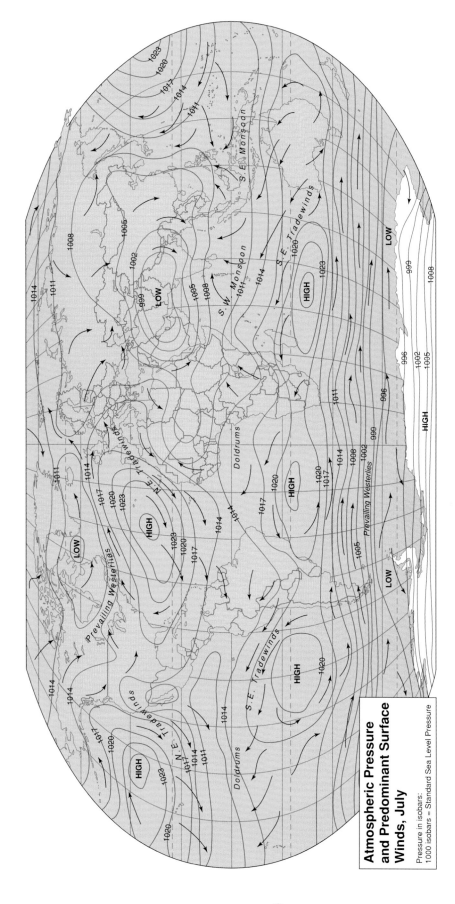

Atmospheric Pressure and Predominant Surface Winds, July

Pressure in isobars:
1000 isobars = Standard Sea Level Pressure

Atmospheric pressure or the density of air is a function largely of air temperature: the colder the air, the denser and heavier it is, hence the higher its pressure; the warmer the air, the lighter and less stable it is, hence the lower its pressure. Global pressure systems are the alternating low and high pressure systems that, from the equator north and south, include: the equatorial low (sometimes called the intertropical convergence) centered on the equator for much of the year; the subtropical highs with their centers near the 30 degrees of north and south latitude; the subpolar lows or polar front centered near the 60th parallel of north and south latitude, and the polar highs near the north and south poles. Air flows from high pressure to low pressure regions and this air flow constitutes the earth's major surface winds such as the tropical tradewinds and the prevailing westerlies. This flow of air is one of the chief mechanisms by which surplus heat energy from the equatorial region is redistributed to higher latitudes. It is also the primary conditioner of the world's major precipitation belts, with rainfall and snowfall associated primarily with lower atmospheric pressure conditions.

Map ⑤ World Climate Regions

Climates of the World

Tropical Moist Climates
- Rain Forest
- Monsoon
- Savanna

Dry Climates
- Warm Desert
- Warm Steppe
- Cool Desert
- Cool Steppe

Midlatitude Climates
- Summer Dry or Mediterranean
- Moist Subtropical
- Marine West Coast
- Cool Forest
- Subarctic

Polar Climates
- Tundra
- Ice Cap

Azonal Climates
- Undifferentiated Highlands

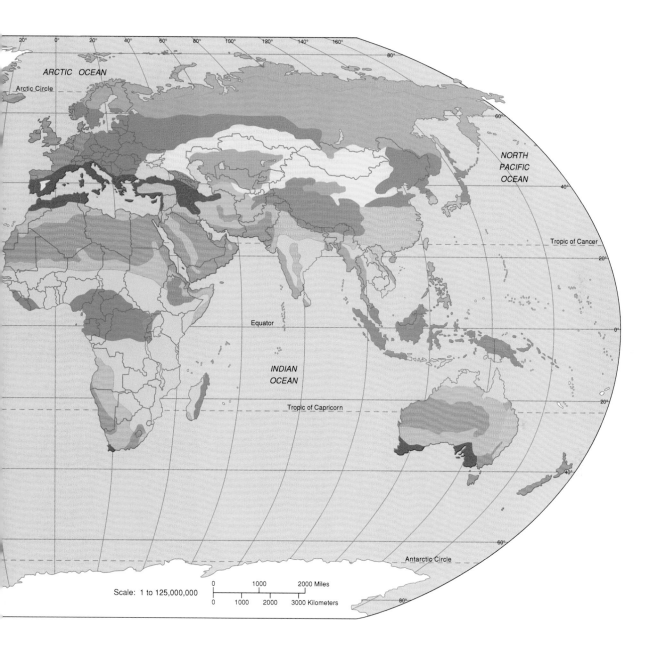

Of the world's many patterns of physical geography, climate or the long-term average of weather conditions such as temperature and precipitation is the most important. It is climate that conditions the distribution of natural vegetation and the types of soils that will exist in an area. Climate also influences the availability of our most crucial resource: water. From an economic standpoint, the world's most important activity is agriculture; no other element of physical geography is more important for agriculture than climate. Ultimately, it is agricultural production that determines where the bulk of human beings live and, therefore, climate is a basic determinant of the distribution of human populations as well. The study of climates or "climatology" is one of the most important branches of physical geography.

Map 6 Vegetation Types

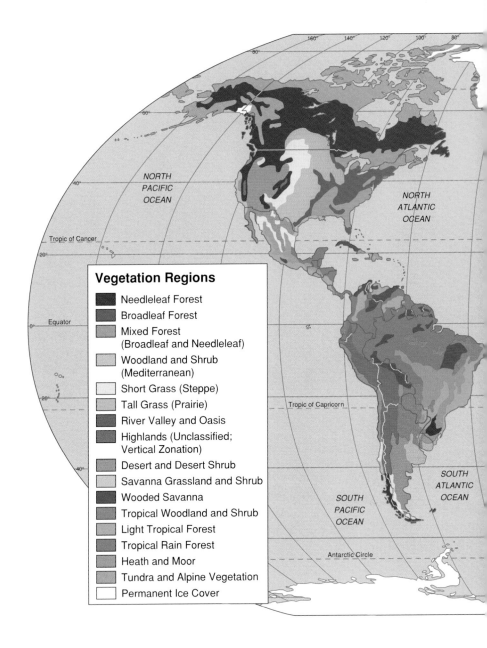

Vegetation Regions

- Needleleaf Forest
- Broadleaf Forest
- Mixed Forest (Broadleaf and Needleleaf)
- Woodland and Shrub (Mediterranean)
- Short Grass (Steppe)
- Tall Grass (Prairie)
- River Valley and Oasis
- Highlands (Unclassified; Vertical Zonation)
- Desert and Desert Shrub
- Savanna Grassland and Shrub
- Wooded Savanna
- Tropical Woodland and Shrub
- Light Tropical Forest
- Tropical Rain Forest
- Heath and Moor
- Tundra and Alpine Vegetation
- Permanent Ice Cover

Vegetation is the most visible consequence of the distribution of temperature and precipitation. The global pattern of vegetative types or "habitat classes" and the global pattern of climate are closely related and make up one of the great global spatial correlations. But not all vegetation types are the consequence of temperature and precipitation or other climatic variables. Many types of vegetation in many areas of the world are the consequence of human activities, particularly the grazing of

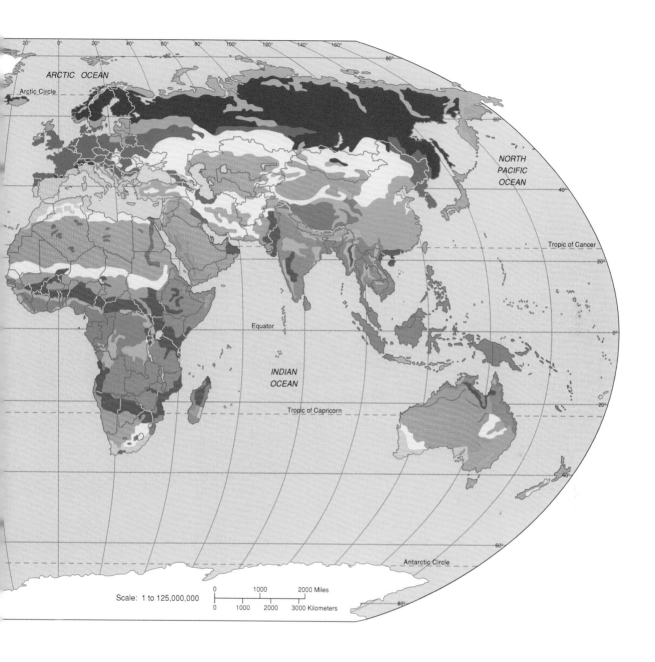

domesticated livestock, burning, and forest clearance. This map shows the pattern of natural or "potential" vegetation, or vegetation as it might be expected to exist without significant human influences, rather than the actual vegetation that results from a combination of environmental and human factors. Physical geographers who are interested in the distribution and geographic patterns of vegetation are "bio-geographers."

Map 7 World Soil Orders

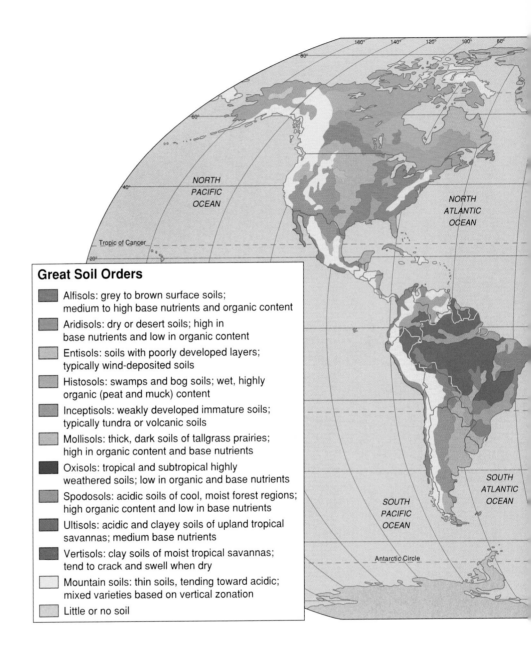

Great Soil Orders

- **Alfisols:** grey to brown surface soils; medium to high base nutrients and organic content
- **Aridisols:** dry or desert soils; high in base nutrients and low in organic content
- **Entisols:** soils with poorly developed layers; typically wind-deposited soils
- **Histosols:** swamps and bog soils; wet, highly organic (peat and muck) content
- **Inceptisols:** weakly developed immature soils; typically tundra or volcanic soils
- **Mollisols:** thick, dark soils of tallgrass prairies; high in organic content and base nutrients
- **Oxisols:** tropical and subtropical highly weathered soils; low in organic and base nutrients
- **Spodosols:** acidic soils of cool, moist forest regions; high organic content and low in base nutrients
- **Ultisols:** acidic and clayey soils of upland tropical savannas; medium base nutrients
- **Vertisols:** clay soils of moist tropical savannas; tend to crack and swell when dry
- **Mountain soils:** thin soils, tending toward acidic; mixed varieties based on vertical zonation
- Little or no soil

The characteristics of soil are one of the three primary physical geographic factors, along with climate and vegetation, that determine the habitability of regions for humans. In particular, soils influence the kinds of agricultural uses to which land is put. Since soils support the plants that are the primary producers of all food in the terrestrial food chain, their characteristics are crucial to the health and stability of ecosystems. Two types of soil are shown on this map: zonal soils, the characteristics of which are based on climatic patterns; and azonal soils, such as alluvial (water-

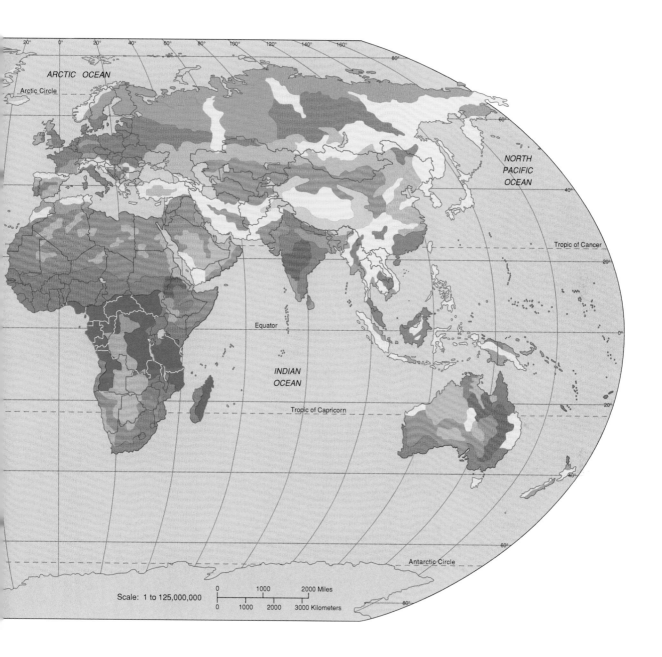

deposited) or aeolian (wind-deposited) soils, the characteristics of which are derived from forces other than climate. However, many of the azonal soils, particularly those dependent upon drainage conditions, appear over areas too small to be readily shown on a map of this scale. Thus, almost none of the world's swamp or bog soils appear on this map. People who study the geographic characteristics of soils are most often "soil scientists," a discipline closely related to that branch of physical geography called "geomorphology."

Map 8 World Ecological Regions

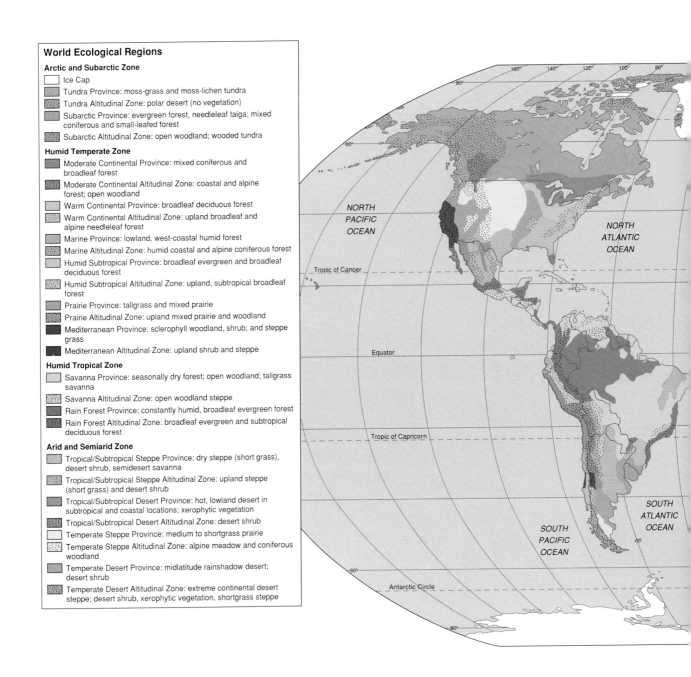

World Ecological Regions

Arctic and Subarctic Zone

Ice Cap

Tundra Province: moss-grass and moss-lichen tundra

Tundra Altitudinal Zone: polar desert (no vegetation)

Subarctic Province: evergreen forest, needleleaf taiga; mixed coniferous and small-leafed forest

Subarctic Altitudinal Zone: open woodland; wooded tundra

Humid Temperate Zone

Moderate Continental Province: mixed coniferous and broadleaf forest

Moderate Continental Altitudinal Zone: coastal and alpine forest; open woodland

Warm Continental Province: broadleaf deciduous forest

Warm Continental Altitudinal Zone: upland broadleaf and alpine needleleaf forest

Marine Province: lowland, west-coastal humid forest

Marine Altitudinal Zone: humid coastal and alpine coniferous forest

Humid Subtropical Province: broadleaf evergreen and broadleaf deciduous forest

Humid Subtropical Altitudinal Zone: upland, subtropical broadleaf forest

Prairie Province: tallgrass and mixed prairie

Prairie Altitudinal Zone: upland mixed prairie and woodland

Mediterranean Province: sclerophyll woodland, shrub, and steppe grass

Mediterranean Altitudinal Zone: upland shrub and steppe

Humid Tropical Zone

Savanna Province: seasonally dry forest; open woodland; tallgrass savanna

Savanna Altitudinal Zone: open woodland steppe

Rain Forest Province: constantly humid, broadleaf evergreen forest

Rain Forest Altitudinal Zone: broadleaf evergreen and subtropical deciduous forest

Arid and Semiarid Zone

Tropical/Subtropical Steppe Province: dry steppe (short grass), desert shrub, semidesert savanna

Tropical/Subtropical Steppe Altitudinal Zone: upland steppe (short grass) and desert shrub

Tropical/Subtropical Desert Province: hot, lowland desert in subtropical and coastal locations; xerophytic vegetation

Tropical/Subtropical Desert Altitudinal Zone: desert shrub

Temperate Steppe Province: medium to shortgrass prairie

Temperate Steppe Altitudinal Zone: alpine meadow and coniferous woodland

Temperate Desert Province: midlatitude rainshadow desert; desert shrub

Temperate Desert Altitudinal Zone: extreme continental desert steppe; desert shrub, xerophytic vegetation, shortgrass steppe

Ecological regions are distinctive areas within which unique sets of organisms and environments are found. We call the study of the relationships between organisms and their environmental surroundings "ecology." Within each of the ecological regions portrayed on the map, a particular combination of vegetation, wildlife, soil, water, climate, and terrain defines that region's habitability, or ability to support life, including human life. Like climate and landforms, ecological relationships are crucial to

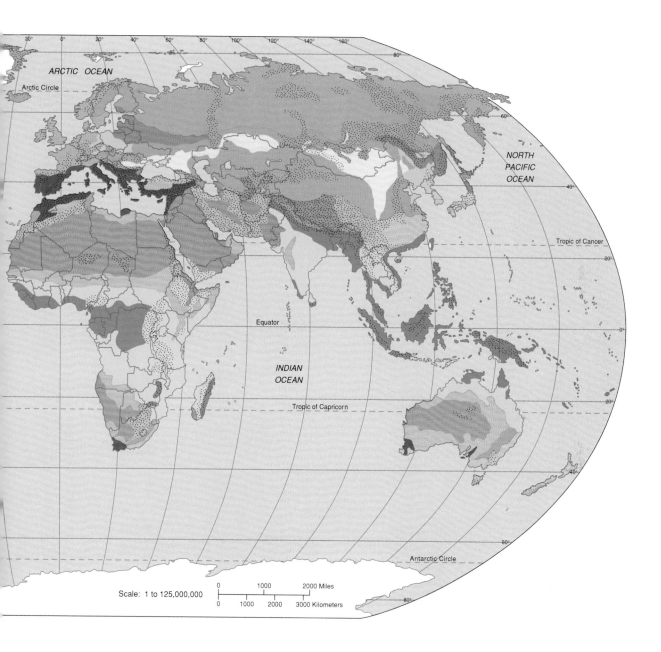

the existence of agriculture, the most basic of our economic activities, and important for many other kinds of economic activity as well. Biogeographers are especially concerned with the concept of ecological regions since such regions so clearly depend upon the geographic distribution of plants and animals in their environmental settings.

Map 9 Plate Tectonics

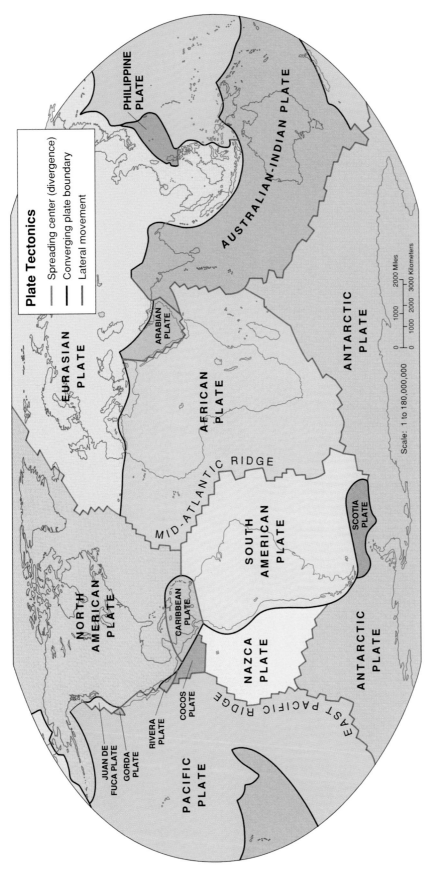

Plate Tectonics

— Spreading center (divergence)
— Converging plate boundary
— Lateral movement

PHILIPPINE PLATE

AUSTRALIAN-INDIAN PLATE

EURASIAN PLATE

ARABIAN PLATE

AFRICAN PLATE

ANTARCTIC PLATE

MID-ATLANTIC RIDGE

SOUTH AMERICAN PLATE

SCOTIA PLATE

NORTH AMERICAN PLATE

CARIBBEAN PLATE

COCOS PLATE

RIVERA PLATE

JUAN DE FUCA PLATE

GORDA PLATE

PACIFIC PLATE

NAZCA PLATE

ANTARCTIC PLATE

EAST PACIFIC RIDGE

Scale: 1 to 180,000,000

0   1000   2000 Miles
0   1000   2000   3000 Kilometers

An understanding of the forces that shape the primary features of the earth's surface—the continents and ocean basins—requires a view of the earth's crust as fragments or "lithospheric plates" that shift position relative to one another. There are three dominant types of plate movement: *convergence*, in which plates move together, compressing former ocean floor or continental rocks together to produce mountain ranges, or producing mountain ranges through volcanic activity if one plate slides beneath another; *divergence*, in which the plates move away from one another, producing rifts in the earth's crust through which molten material wells up to produce new sea floors and mid-oceanic ridges; and *lateral shift*, in which plates move horizontally relative to one another, causing significant earthquake activity. All the major forms of these types of shifts are extremely slow and take place over long periods of geologic time. The movement of crustal plates, or what is known as "plate tectonics," is responsible for the present shape and location of the continents but is also the driving force behind some much shorter-term earth phenomena like earthquakes and volcanoes. A comparison of the map of plates with maps of hazards and terrain will reveal some interesting relationships.

Map 10 World Topography

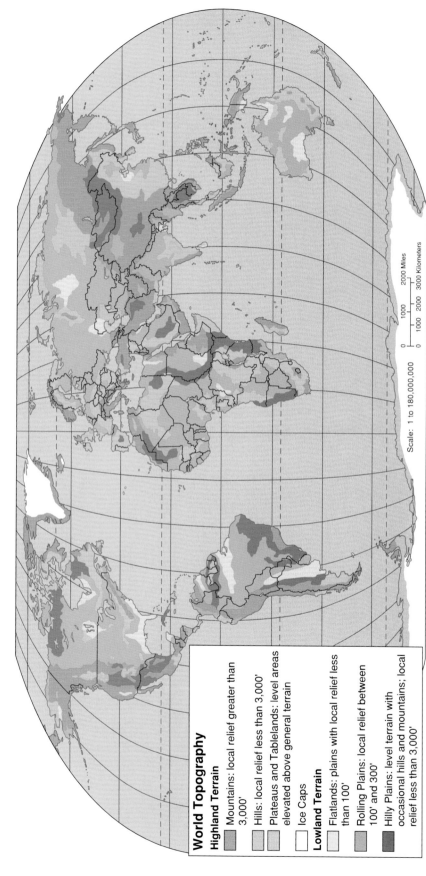

World Topography

Highland Terrain

- Mountains: local relief greater than 3,000'
- Hills: local relief less than 3,000'
- Plateaus and Tablelands: level areas elevated above general terrain
- Ice Caps

Lowland Terrain

- Flatlands: plains with local relief less than 100'
- Rolling Plains: local relief between 100' and 300'
- Hilly Plains: level terrain with occasional hills and mountains; local relief less than 3,000'

Scale: 1 to 180,000,000

0   1000   2000 Miles
0   1000   2000   3000 Kilometers

Topography or terrain, also called "landforms," is second only to climate as a conditioner of human activity, particularly agriculture but also the location of cities and industry. A comparison of this map of mountains, valleys, plains, plateaus, and other features of the earth's surface with a map of land use (Map 13) shows that most of the world's productive agricultural zones are located in lowland and relatively level regions. Where large regions of agricultural productivity are found, we also tend to find urban concentrations and, with cities, we find industry. There is also a good spatial correlation between the map of topography and the map showing the distribution and density of the human population (Map 12). Normally the world's major landforms are the result of extremely gradual primary geologic activity such as the long-term movement of crustal plates. This activity occurs over hundreds of millions of years. Also important is the more rapid (but still slow by human standards) geomorphological or erosional activity of water, wind, glacial ice, and waves, tides, and currents. Some landforms may be produced by abrupt or "cataclysmic" events such as a major volcanic eruption or a meteor strike, but such events are relatively rare and their effects are usually too minor to show up on a map of this scale. The study of the processes that shape topography is known as "geomorphology" and is an important branch of physical geography.

Map 11 World Mineral Fuels and Critical Metals

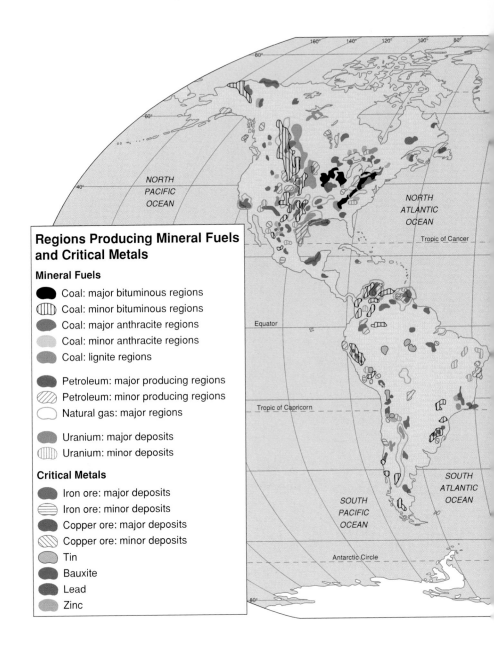

Regions Producing Mineral Fuels and Critical Metals

Mineral Fuels

- Coal: major bituminous regions
- Coal: minor bituminous regions
- Coal: major anthracite regions
- Coal: minor anthracite regions
- Coal: lignite regions

- Petroleum: major producing regions
- Petroleum: minor producing regions
- Natural gas: major regions

- Uranium: major deposits
- Uranium: minor deposits

Critical Metals

- Iron ore: major deposits
- Iron ore: minor deposits
- Copper ore: major deposits
- Copper ore: minor deposits
- Tin
- Bauxite
- Lead
- Zinc

The extraction and transportation of mineral fuels and critical metal ores rank with agriculture and forestry as "primary" human activities that impact on the environment at a global scale. Nearly all of the most highly publicized environmental disasters of recent decades—including the Prince William Sound oil spill, the Chernobyl nuclear accident, and the Persian Gulf War oil spills—have involved mineral fuels that were being stored, transported, or used. And the continuing extraction of critical mineral fuels and metal ores produces high levels of atmospheric, soil, and water pollution. The location of fuels and metal ores tells us a great deal about where environmental degradation is likely to be occurring or to occur in the future. But the location of these materials also tells us much about economic development and

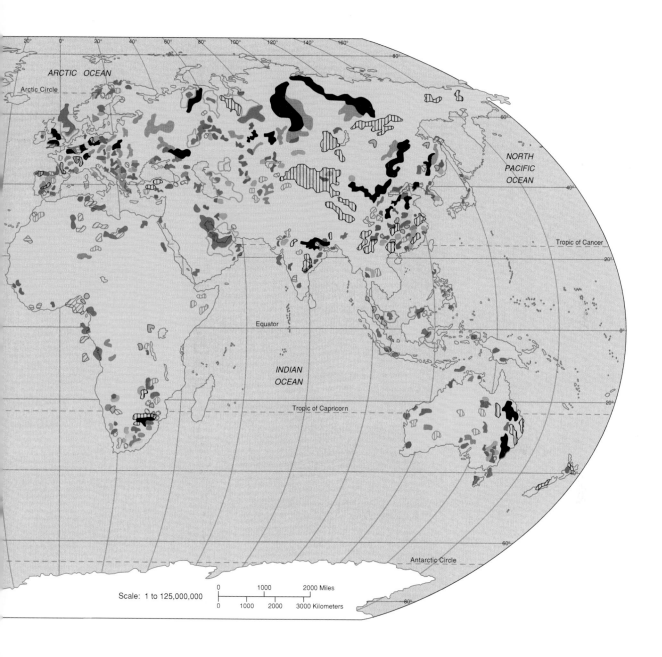

well-being. Generally, those countries and regions that rank higher on the scale of development are those with more abundant mineral resources. Among the developing countries, those with significant deposits of mineral fuels or critical metal ores stand the greatest chance of building the industries that can help them make the transition to the higher standards of living and reduced population growth that are among the identifiers of more economically developed areas. The location and significance of mineral fuels and critical metals is studied both by physical geographers specializing in minerals and by economic geographers interested in the distribution of primary economic activities.

Map 12   World Natural Hazards

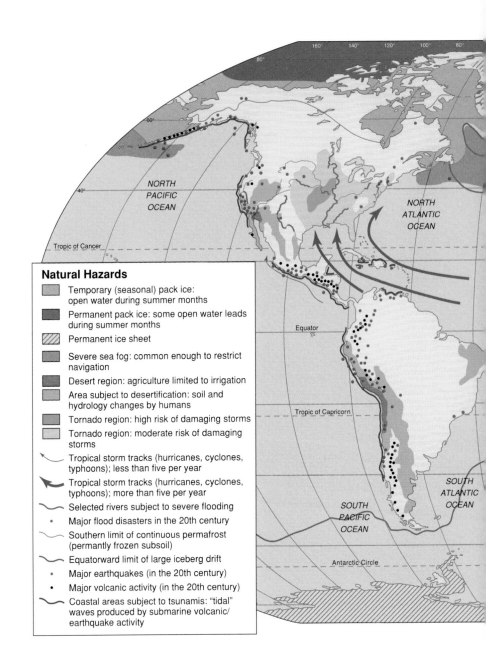

Natural Hazards

Temporary (seasonal) pack ice: open water during summer months

Permanent pack ice: some open water leads during summer months

Permanent ice sheet

Severe sea fog: common enough to restrict navigation

Desert region: agriculture limited to irrigation

Area subject to desertification: soil and hydrology changes by humans

Tornado region: high risk of damaging storms

Tornado region: moderate risk of damaging storms

Tropical storm tracks (hurricanes, cyclones, typhoons); less than five per year

Tropical storm tracks (hurricanes, cyclones, typhoons); more than five per year

Selected rivers subject to severe flooding

Major flood disasters in the 20th century

Southern limit of continuous permafrost (permantly frozen subsoil)

Equatorward limit of large iceberg drift

Major earthquakes (in the 20th century)

Major volcanic activity (in the 20th century)

Coastal areas subject to tsunamis: "tidal" waves produced by submarine volcanic/earthquake activity

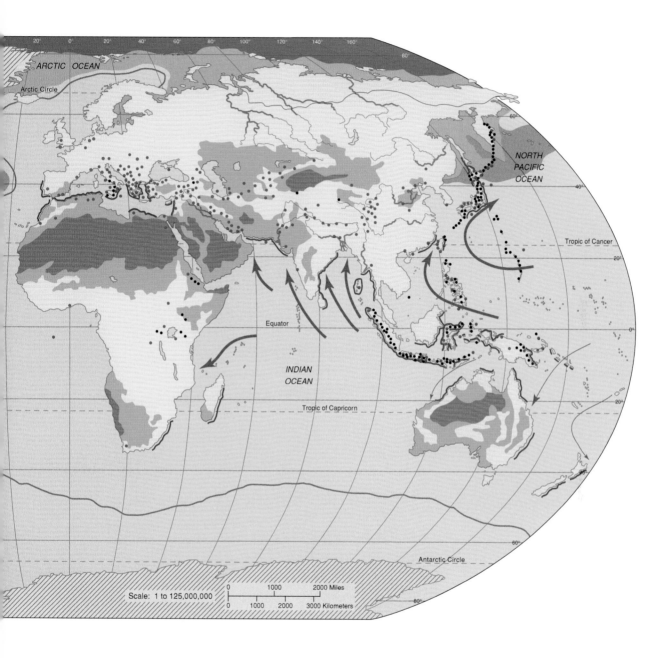

Unlike other elements of physical geography, most natural hazards are unpredictable. However, there are certain regions where the *probability* of the occurrence of a particular natural hazard is high. This map shows regions affected by major natural hazards at rates that are higher than the global norm. The presence of persistent natural hazards may influence the types of modifications that people make in the environment and certainly influence the styles of housing and other elements of cultural geography. Natural hazards may also undermine the utility of an area for economic purposes and some scholars suggest that regions of environmental instability may be regions of political instability as well. The study of natural hazards has become an important activity for "resource geographers" whose areas of interest overlap both human and physical fields of geography.

Part II

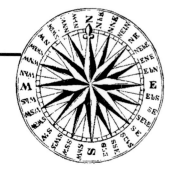

The Human Environment: Global Patterns

Map 13 World Population Density

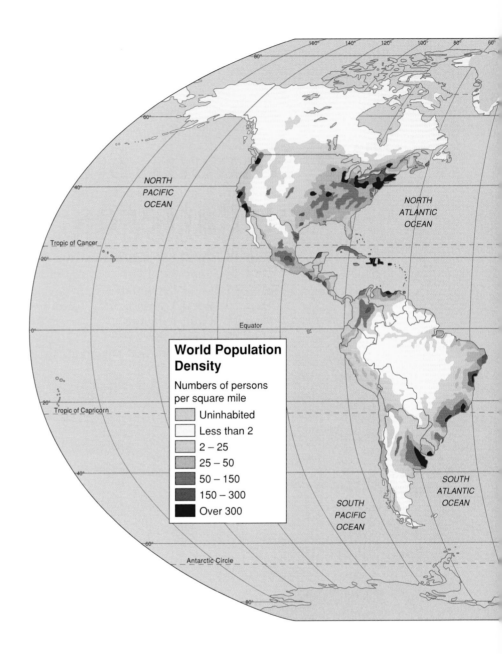

World Population Density

Numbers of persons per square mile

	Uninhabited
	Less than 2
	2 – 25
	25 – 50
	50 – 150
	150 – 300
	Over 300

No feature of human activity is more reflective of geographic relationships than where people live. In the areas of densest populations, a mixture of natural and human factors has combined to allow maximum food production, maximum urbanization, and maximum centralization of economic activities. Three great concentrations of human population appear on the map—East Asia, South Asia, and Europe—with a fourth, lesser concentration in eastern North America. While population growth is relatively slow in three of these population clusters, in the fourth— South Asia—growth is still rapid and South Asia is expected to become even more densely populated in the early years of the twenty-first century while density of the other three regions is expected to remain about as it now appears. In Europe and

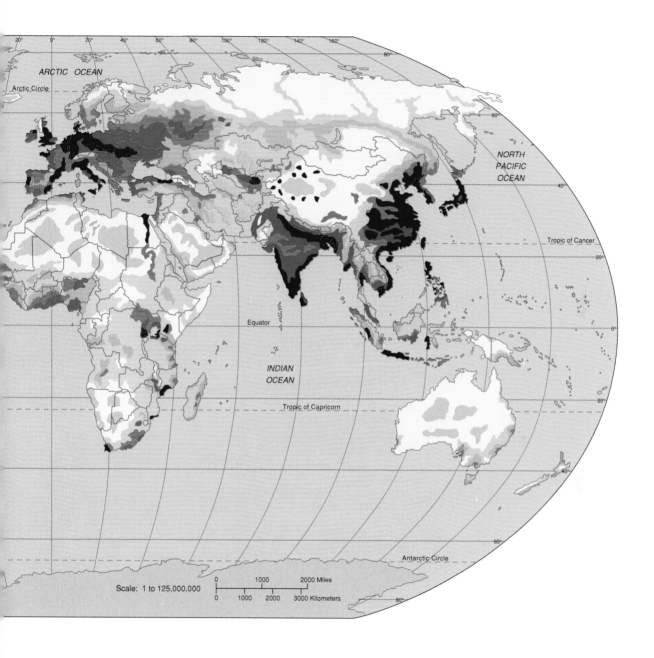

North America, the relatively stable population growth rates are the result of economic development that has caused population growth to level off within the last century. In East Asia, the growth rates have also begun to decline. In the case of Japan, Taiwan, the Koreas, and other more highly developed nations of the Pacific Rim, the reduced growth is the result of economic development. In China, at least until recently, lowered population growth rates have resulted from strict family planning. The areas of future high density of population, in addition to those already existing, are likely to be in Middle and South America and in Central Africa, where population growth rates are well above the world average.

Map 14 World Land Use

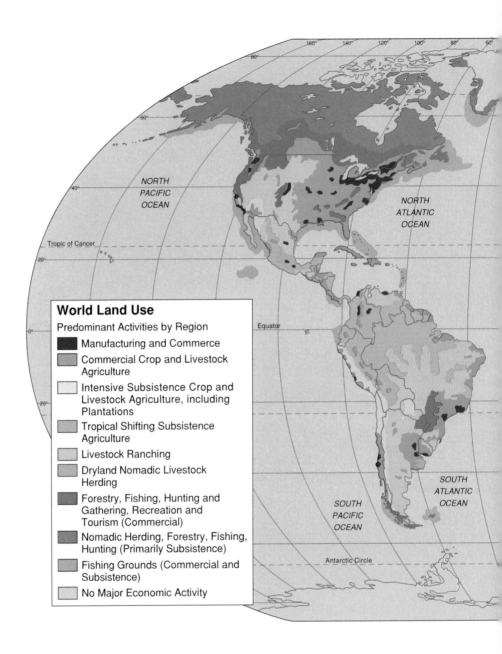

World Land Use

Predominant Activities by Region

- Manufacturing and Commerce
- Commercial Crop and Livestock Agriculture
- Intensive Subsistence Crop and Livestock Agriculture, including Plantations
- Tropical Shifting Subsistence Agriculture
- Livestock Ranching
- Dryland Nomadic Livestock Herding
- Forestry, Fishing, Hunting and Gathering, Recreation and Tourism (Commercial)
- Nomadic Herding, Forestry, Fishing, Hunting (Primarily Subsistence)
- Fishing Grounds (Commercial and Subsistence)
- No Major Economic Activity

Land uses can be categorized as lying somewhere on a scale between *extensive* uses in which human activities are dispersed over relatively large areas and *intensive* uses in which human activities are concentrated in relatively small areas. Many of the most important land use patterns of the world (such as urbanization, industry, mining, or transportation) are intensive and therefore relatively small in area and not easily seen on maps of this scale. Hence, even in the areas identified as "Manufacturing and Commerce" on the map there are many land uses that are not strictly industrial or commercial in nature and, in fact, more extensive land uses (farming, residential, open space) may actually cover more ground than the intensive industrial

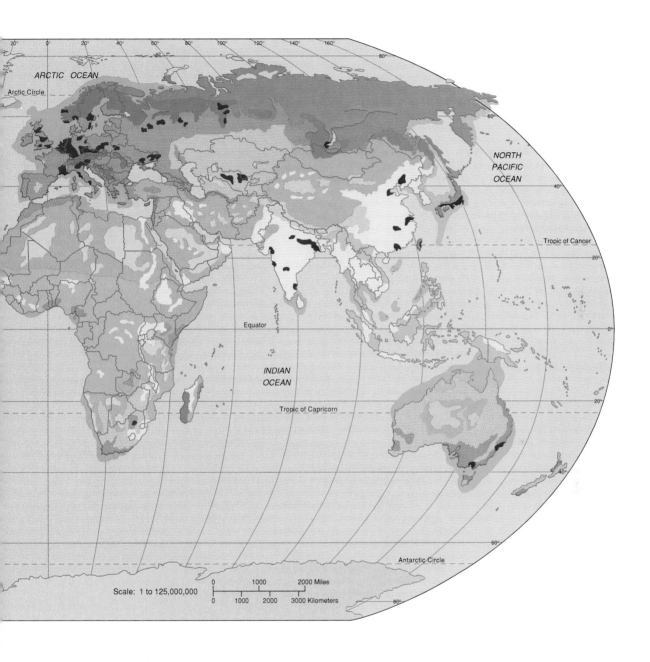

or commercial activities. On the other hand, the more extensive land uses, like agriculture and forestry, tend to dominate the areas in which they are found. Thus, primary economic activities such as agriculture and forestry tend to dominate the world map of land use because of their extensive character. Much of this map is, therefore, a map that shows the global variations in agricultural patterns. Note, among other things, the differences between land use patterns in the more developed countries of the temperate zones and the less developed countries of the tropics.

Map 15 World Urbanization

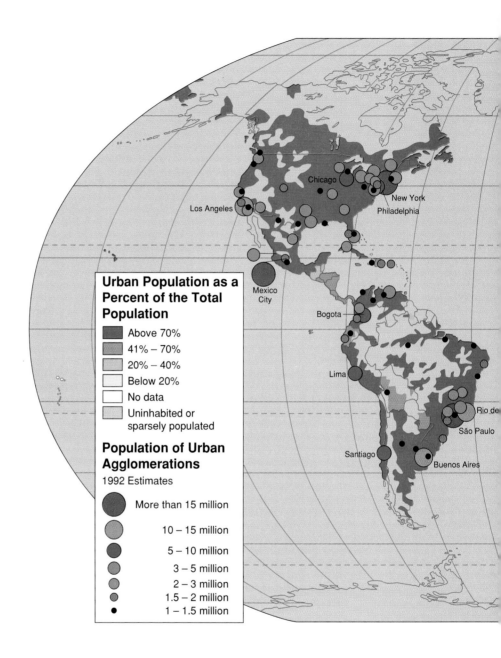

Urban Population as a Percent of the Total Population

- Above 70%
- 41% – 70%
- 20% – 40%
- Below 20%
- No data
- Uninhabited or sparsely populated

Population of Urban Agglomerations

1992 Estimates

- More than 15 million
- 10 – 15 million
- 5 – 10 million
- 3 – 5 million
- 2 – 3 million
- 1.5 – 2 million
- 1 – 1.5 million

Chicago
New York
Philadelphia
Los Angeles
Mexico City
Bogota
Lima
Santiago
Buenos Aires
Rio de
São Paulo

The degree to which a region's population is concentrated in urban areas is a major indicator of a number of things: the potential for environmental impact, the level of economic development, and the problems associated with human concentrations. Urban dwellers are rapidly becoming the norm among the world's people and rates of urbanization are increasing worldwide, with the greatest increases in urbanization taking place in developing regions. Whether in developed or developing countries, those who live in cities exert an influence on the environment, politics, economics, and social systems that go far beyond the confines of the city itself. Acting as the

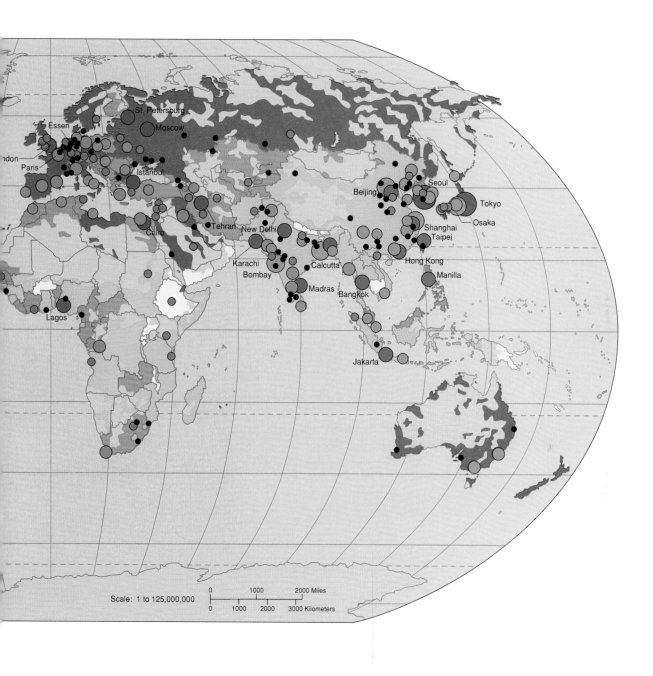

focal points for the flow of goods and ideas, cities draw resources and people not just from their immediate hinterland but from the entire world. This process creates far-reaching impacts as resources are extracted, converted through industrial processes, and transported over great distances to metropolitan regions, and as ideas spread or *diffuse* along with the movements of people to cities and the flow of communication from them. The significance of urbanization can be most clearly seen, perhaps, in North America where, in spite of vast areas of relatively unpopulated land, well over 90% of the population lives in urban areas.

Map 16 World Transportation Patterns

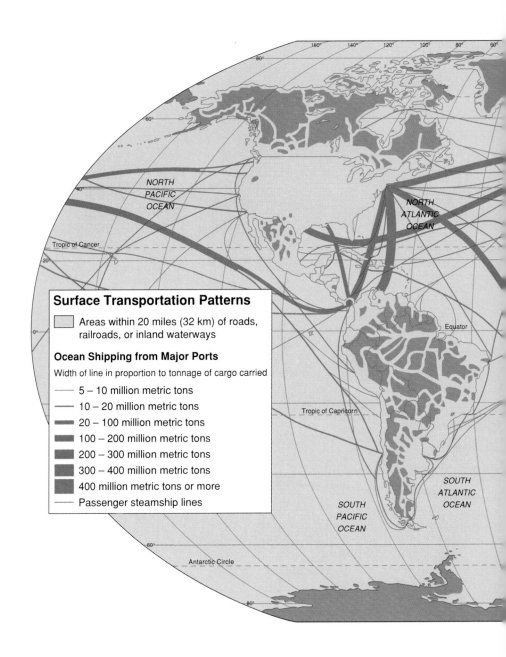

Surface Transportation Patterns

Areas within 20 miles (32 km) of roads, railroads, or inland waterways

Ocean Shipping from Major Ports

Width of line in proportion to tonnage of cargo carried

— 5 – 10 million metric tons

— 10 – 20 million metric tons

— 20 – 100 million metric tons

— 100 – 200 million metric tons

— 200 – 300 million metric tons

— 300 – 400 million metric tons

— 400 million metric tons or more

— Passenger steamship lines

As a form of land use, transportation is second only to agriculture in its coverage of the earth's surface and is one of the clearest examples in the human world of a *network,* a linked system of lines allowing flows from one place to another. The global transportation network and its related communication web is responsible for most of the *spatial interaction* or movement of goods, people, and ideas between places. As the chief mechanism of spatial interaction, transportation is linked firmly with the concept of a shrinking world and the development of a global community

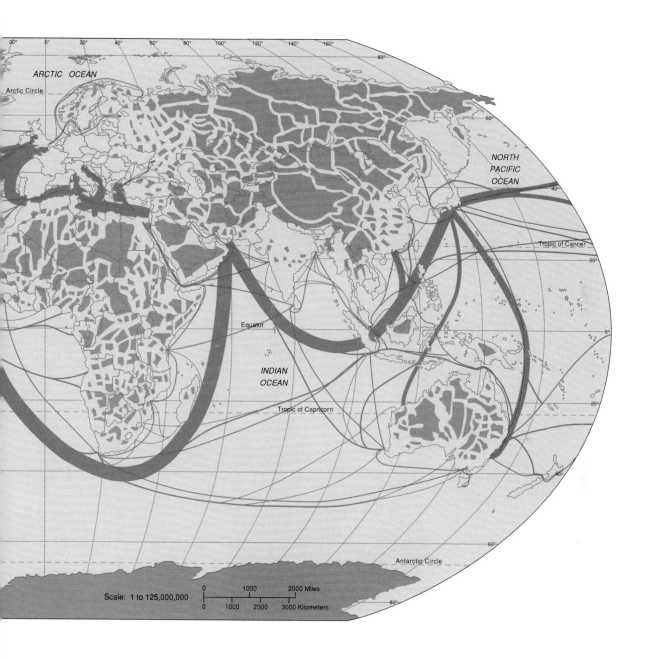

and economy. Because transportation systems require significant modification of the earth's surface, transportation is also responsible for massive alterations in the quantity and quality of water, for major soil degradations and erosion, and (indirectly) for the air pollution that emanates from vehicles utilizing the transportation system. In addition, as improved transportation technology draws together places on the earth that were formerly remote, it allows people to impact environments a great distance away from where they live.

Map 17 World Religions

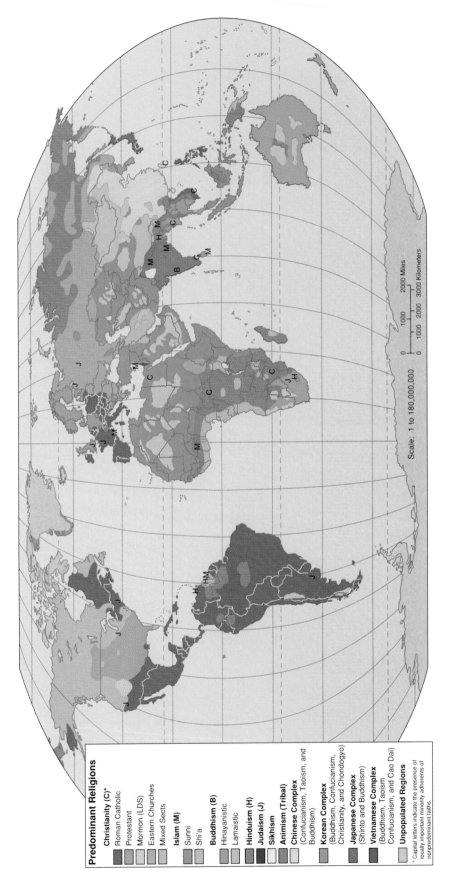

Predominant Religions

Christianity (C)*
- Roman Catholic
- Protestant
- Mormon (LDS)
- Eastern Churches
- Mixed Sects

Islam (M)
- Sunni
- Shi'a

Buddhism (B)
- Hinayanistic
- Lamaistic

Hinduism (H)

Judaism (J)

Sikhism

Animism (Tribal)

Chinese Complex
(Confucianism, Taoism, and Buddhism)

Korean Complex
(Buddhism, Confucianism, Christianity, and Chondogyo)

Japanese Complex
(Shinto and Buddhism)

Vietnamese Complex
(Buddhism, Taoism Confucianism, and Cao Dai)

Unpopulated Regions

* Capital letters indicate the presence of locally important minority adherents of nonpredominant faiths.

Scale: 1 to 180,000,000

0   1000   2000 Miles

0   1000   2000   3000 Kilometers

Religious adherence is one of the fundamental defining characteristics of human *culture*, the style of life adopted by a people and passed from one generation to the next. Because of the importance of religion for culture, a depiction of the spatial distribution of religions is as close as we can come to a map of cultural patterns. More than just a set of behavioral patterns having to do with worship and ceremony, religion is a vital conditioner of the ways that people deal with one another, with their institutions, and with the environments they occupy. In many areas of the world, the ways in which people make a living, the patterns of occupation that they create on the land, and the impacts that they make on ecosystems are the direct consequences of their adherence to a religious faith. An examination of the map in the context of international and intranational conflict will also show that tension between countries and the internal stability of states is also a function of the spatial distribution of religion.

Map 18 World Languages

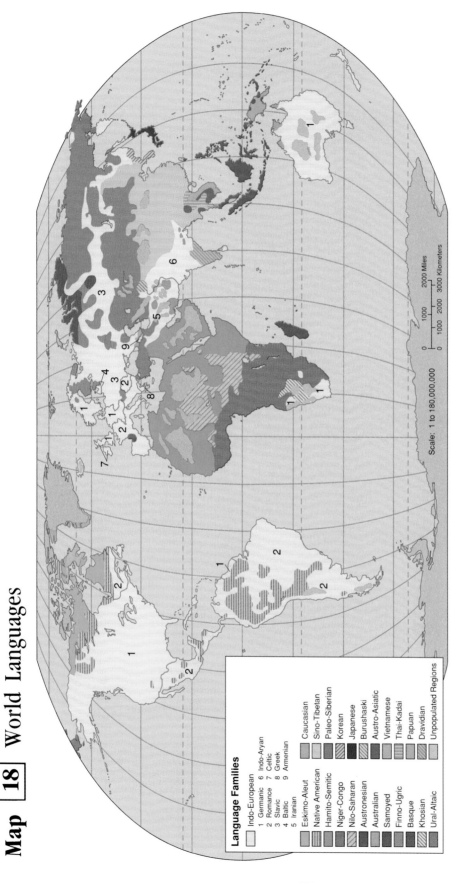

Language Families

Indo-European

1 Germanic   6 Indo-Aryan
2 Romance   7 Celtic
3 Slavic   8 Greek
4 Baltic   9 Armenian
5 Iranian

- Eskimo-Aleut
- Native American
- Hamito-Semitic
- Niger-Congo
- Nilo-Saharan
- Austronesian
- Australian
- Samoyed
- Finno-Ugric
- Basque
- Khosian
- Ural-Altaic
- Caucasian
- Sino-Tibetan
- Paleo-Siberian
- Korean
- Japanese
- Burushaski
- Austro-Asiatic
- Vietnamese
- Thai-Kadai
- Papuan
- Dravidian
- Unpopulated Regions

Scale: 1 to 180,000,000

0   1000   2000   2000 Miles

0   1000   2000   3000 Kilometers

Language, like religion, is an important identifying characteristic of culture. Indeed, it is perhaps the most durable of all those identifying characteristics or *cultural traits*: language, religion, institutions, material technologies, and ways of making a living. After centuries of exposure to other languages or even conquest by speakers of other languages, the speakers of a specific tongue will often retain their own linguistic identity. As a geographic element, language helps us to locate areas of potential conflict, particularly in regions where two or more languages overlap. Many, if not most, of the world's conflict zones are also areas of linguistic diversity and knowing the distribution of languages helps us to understand some of the reasons behind important current events: for example, linguistic identity differences played

an important part in the disintegration of the Soviet Union in the early 1990s; and in areas emerging from recent colonial rule, such as Africa, the participants in conflicts over territory and power are often defined in terms of linguistic groups. Language distributions also help us to comprehend the nature of the human past by providing clues that enable us to chart the course of human migrations, as shown in the distribution of Indo-European, Austronesian, or Hamito-Semitic languages. Finally, because languages have a great deal to do with the way people perceive and understand the world around them, linguistic patterns help to explain the global variations in the ways that people interact with their environments.

Map 19 World External Migrations in Modern Times

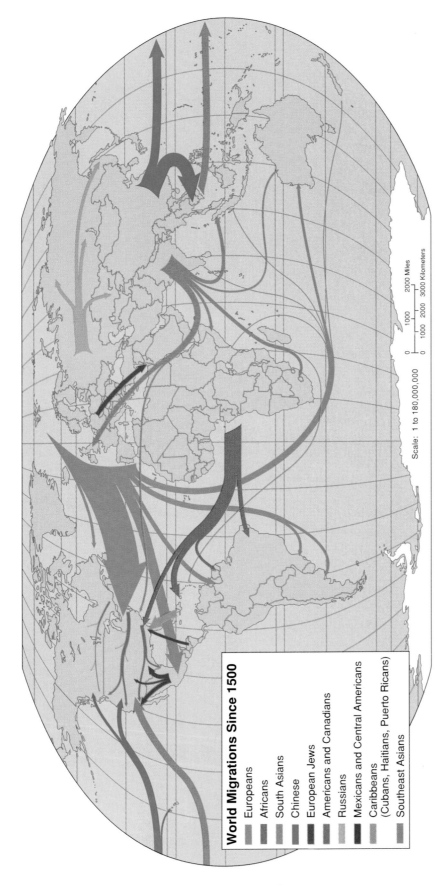

World Migrations Since 1500

- Europeans
- Africans
- South Asians
- Chinese
- European Jews
- Americans and Canadians
- Russians
- Mexicans and Central Americans
- Caribbeans (Cubans, Haitians, Puerto Ricans)
- Southeast Asians

Scale: 1 to 180,000,000

0   1000   2000 Miles

0   1000   2000   3000 Kilometers

Migration has had a significant effect on world geography, contributing to cultural change and development, to the diffusion of ideas and innovations, and to the complex mixture of people and cultures found in the world today. *Internal migration* occurs within the boundaries of a country; *external migration* is movement from one country or region to another. Over the last 50 years, the most important migrations in the world have been internal, largely the rural-to-urban migration that has been responsible for the recent rise of global urbanization. Prior to the mid-20th century, 3 types of external migrations were most important: *voluntary*, most often in search of better economic conditions and opportunities; *involuntary or forced*, involving people who have been driven from their homelands by war, political unrest, or environmental disasters, or who have been transported as slaves or prisoners; and *imposed*, not entirely forced but which conditions make highly advisable. Human migrations in recorded history have been responsible for major changes in the patterns of languages, religions, ethnic composition, and economies. Particularly during the last 500 years, migrations of both the voluntary and involuntary or forced type have literally reshaped the human face of the earth.

Part III

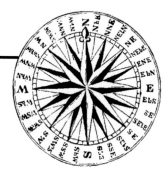

Global Demographic Patterns

Map 20 Population Growth Rates

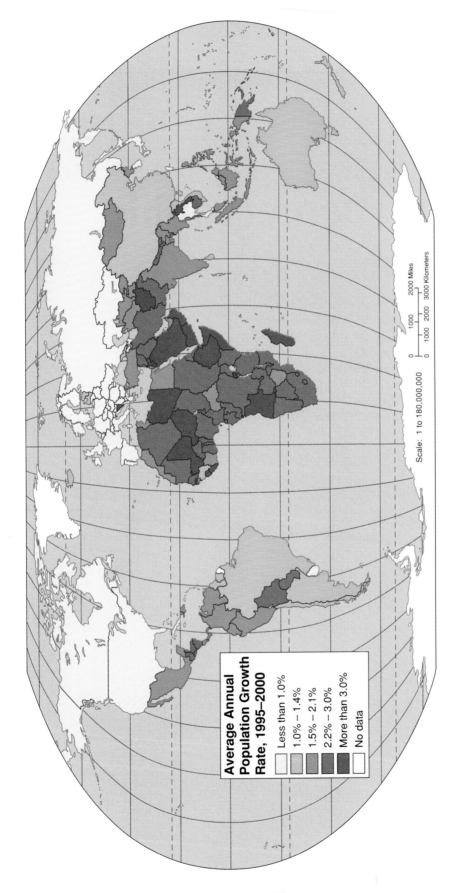

Average Annual Population Growth Rate, 1995–2000

- Less than 1.0%
- 1.0% – 1.4%
- 1.5% – 2.1%
- 2.2% – 3.0%
- More than 3.0%
- No data

Scale: 1 to 180,000,000

0   1000   2000   Miles

0   1000   2000   3000 Kilometers

Of all the statistical measurements of human population, that of the rate of population growth is the most important. The growth rate of a population is a combination of natural change (births and deaths), in-migration, and out-migration; it is obtained by adding the number of births to the number of immigrants during a year and subtracting from that total the sum of deaths and emigrants for the same year. For a specific country, this figure will determine many things about the country's future ability to feed, house, educate, and provide medical services to its citizens. Some of the countries with the largest populations (such as India) also have high growth rates. Since these countries tend to be in developing regions, the combination of high population and high growth rates poses special problems for continuing economic development and carries heightened risks of environmental degradation. Many people believe that the rapidly expanding world population is a potential crisis that may cause environmental and human disaster by the middle of the twenty-first century.

Map 21 Total Fertility Rate

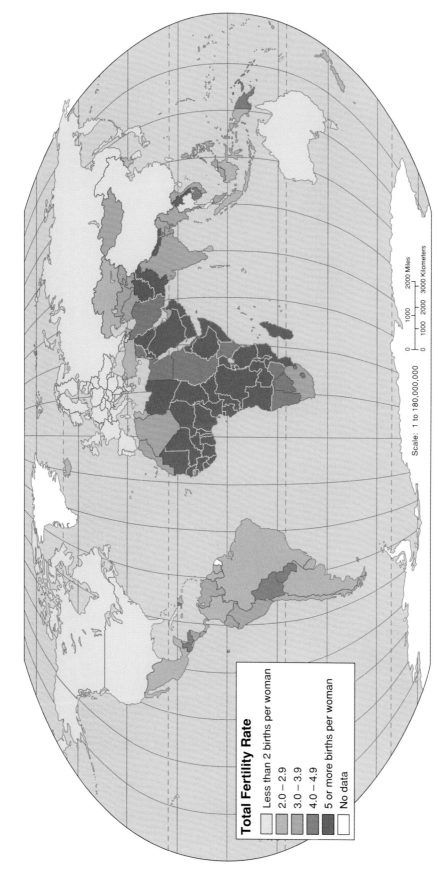

Total Fertility Rate
- Less than 2 births per woman
- 2.0 – 2.9
- 3.0 – 3.9
- 4.0 – 4.9
- 5 or more births per woman
- No data

Scale: 1 to 180,000,000

The fertility rate measures the number of children that a woman is expected to bear during her lifetime, based on the age-specific fertility figures of women between 15 and 40 (the normal childbearing years). While fertility rates tell us a great deal about present population growth, with high fertility rates indicating high population growth rates, they are also indicative of potential or projected growth. A country whose women can be expected to bear many children is a country with enormous potential for population growth in the future. Given present fertility rates, for example, the number of offspring from the average German woman over the next three generations (the total number of children, grandchildren, and great-grandchildren)

will be 7. During the same three generations, the average American woman will have a total of 17 children, grandchildren, and great-grandchildren. But during this time, assuming that present fertility rates are maintained, the average woman in sub-Saharan Africa will have 258 children, grandchildren, and great-grandchildren. You might be interested in working out some potential population growth rates over two or three generations, using the data as presented on the map.

Note: Total fertility rate figures are based on yearly age-specific fertility rate estimates from 1990–2000.

Map 22 Infant Mortality Rate

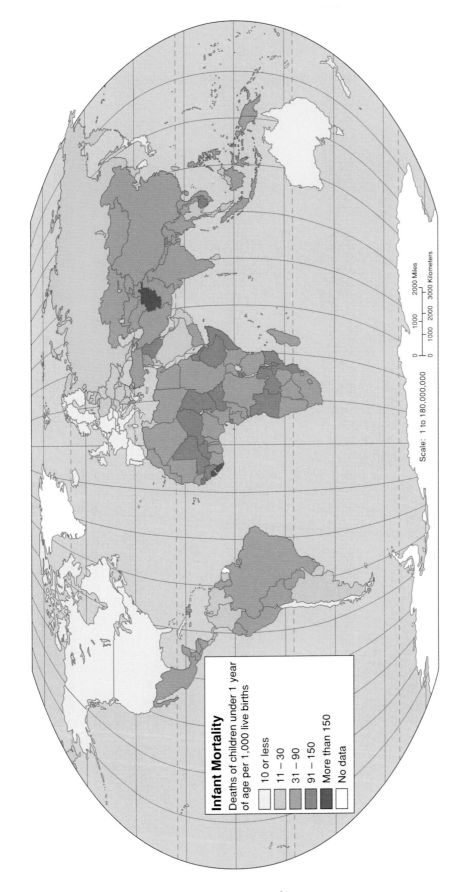

Infant Mortality

Deaths of children under 1 year of age per 1,000 live births

- 10 or less
- 11 – 30
- 31 – 90
- 91 – 150
- More than 150
- No data

Scale: 1 to 180,000,000

0   1000   2000 Miles

0   1000   2000   3000 Kilometers

Infant mortality rates are calculated by dividing the number of children born in a given year who die before their first birthday by the total number of children born that year and then multiplying by 1,000; this shows how many infants have died for every 1,000 births. Infant mortality rates are prime indicators of economic development. In highly developed economies, with advanced medical technologies, sufficient diets, and adequate public sanitation, infant mortality rates tend to be quite low. By contrast, in less developed countries, with the disadvantages of poor diet, limited access to medical technology, and the other problems of poverty, infant mortality rates

tend to be high. Although worldwide infant mortality has decreased significantly during the last 2 decades, many regions of the world still experience infant mortality above the 10 percent level (100 deaths per 1,000 live births). Such infant mortality rates represent not only human tragedy at its most basic level, but also are powerful inhibiting factors for the future of human development. Comparing infant mortality rates in the midlatitudes and the tropics shows that children in most African countries are more than 10 times as likely to die within a year of birth as children in European countries.

Map 23 Average Life Expectancy at Birth

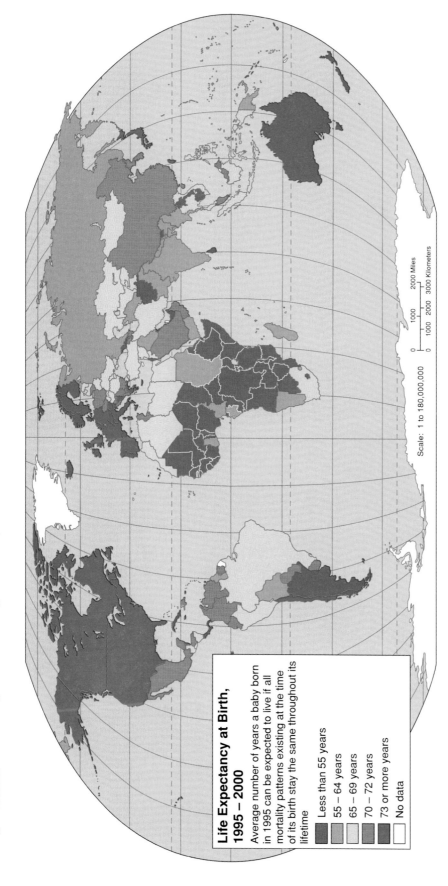

Life Expectancy at Birth, 1995 – 2000

Average number of years a baby born in 1995 can be expected to live if all mortality patterns existing at the time of its birth stay the same throughout its lifetime

- Less than 55 years
- 55 – 64 years
- 65 – 69 years
- 70 – 72 years
- 73 or more years
- No data

Scale: 1 to 180,000,000

| 0 | 1000 | 2000 Miles |

| 0 | 1000 | 2000 | 3000 Kilometers |

Average life expectancy at birth is a measure of the average longevity of the population of a country. Like all average measures, it is distorted by extremes. For example, a country with a high mortality rate among children will have a low average life expectancy. Thus, an average life expectancy of 45 years does not mean that everyone can be expected to die at the age of 45. More normally, what the figure means is that a substantial number of children die between birth and 5 years of age, thus reducing the average life expectancy for the entire population. In spite of the dangers inherent in misinterpreting the data, average life expectancy (along with infant mortality and several other measures) is a valid way of judging the relative health of a population. It reflects the nature of the health care system, public sanitation and disease control, nutrition, and a number of other key human need indicators. As such, it is a measure of well-being that is significant in indicating economic development.

Map 24 Population by Age Group

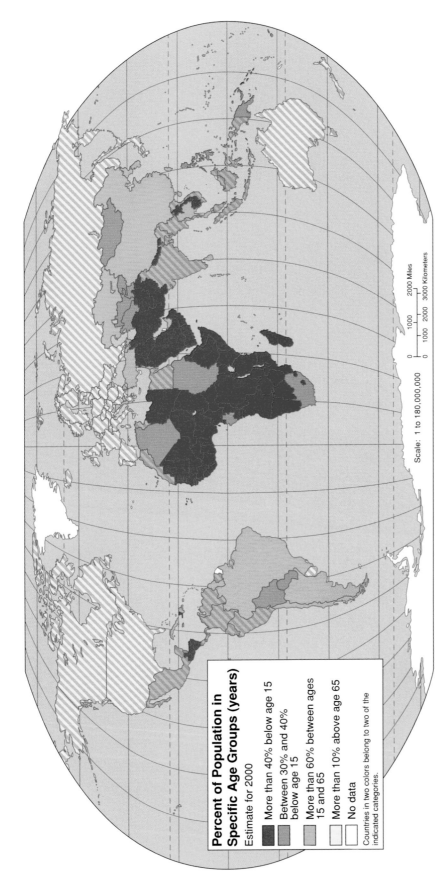

Percent of Population in Specific Age Groups (years)

Estimate for 2000

- More than 40% below age 15
- Between 30% and 40% below age 15
- More than 60% between ages 15 and 65
- More than 10% above age 65
- No data

Countries in two colors belong to two of the indicated categories.

Scale: 1 to 180,000,000

0   1000   2000 Miles

0   1000   2000   3000 Kilometers

Of all the measurements that illustrate the dynamics of a population, age distribution may be the most significant, particularly when viewed in combination with average growth rates. The particular relevance of age distribution is that it tells us what to expect from a population in terms of growth over the next generation. If, for example, approximately 40–50 percent of a population is below the age of 15, that suggests that in the next generation about one-quarter of the total population will be women of childbearing age. When age distribution is combined with fertility rates (the average number of children born per woman in a population), an especially valid measure-ment of future growth potential may be derived. A simple example: Nigeria, with a 1995 population of 127 million, has 47 percent of its population below the age of 15 and a fertility rate of 6.4; the United States, with a 1995 population of 263 million, has 22 percent of its population below the age of 15 and a fertility rate of 2.1. During the period in which those women presently under the age of 15 are in their childbearing years, Nigeria can be expected to add a total of approximately 197 million persons to its total population. Over the same period, the United States can be expected to add only 61 million.

Map 25 World Daily Per Capita Food Supply

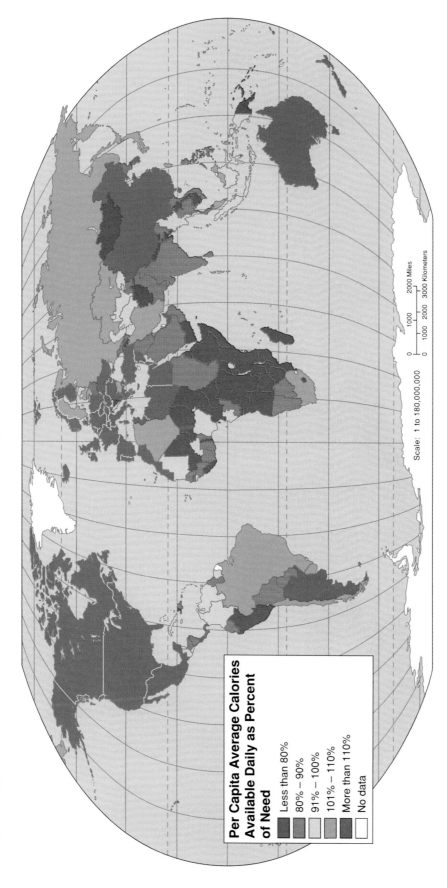

Per Capita Average Calories Available Daily as Percent of Need

- Less than 80%
- 80% – 90%
- 91% – 100%
- 101% – 110%
- More than 110%
- No data

Scale: 1 to 180,000,000

0   1000   2000 Miles
0 1000 2000 3000 Kilometers

The data shown on this map indicate the presence or absence of critical food shortages. While they do not necessarily indicate the presence of starvation or famine, they certainly do indicate potential problem areas for the next decade. The measurements are in calories from *all* food sources: domestic production, international trade, drawdown on stocks or food reserves, and direct foreign contributions or aid. The quantity of calories available is that amount, estimated by the UN's Food and Agriculture Organization (FAO), that actually reaches consumers. The calories actually consumed may be lower than the figures shown, depending on how much is lost in a variety of ways: in home storage (to pests such as rats and mice), in preparation and cooking, through consumption by pets and domestic animals, and as discarded foods, for example. The former practice in such maps was to evaluate available calories as a percent of "need" or minimum daily requirements to maintain health. Such a statistical measure was virtually impossible to standardize for the variety of human types in the world and for such variables as age and sex distribution. A newer form of measure—available calories as a percent of the world average available—eliminates many of the problems of the former set of numbers while still maintaining a good relative picture of global hunger.

Map 26 Illiteracy Rate

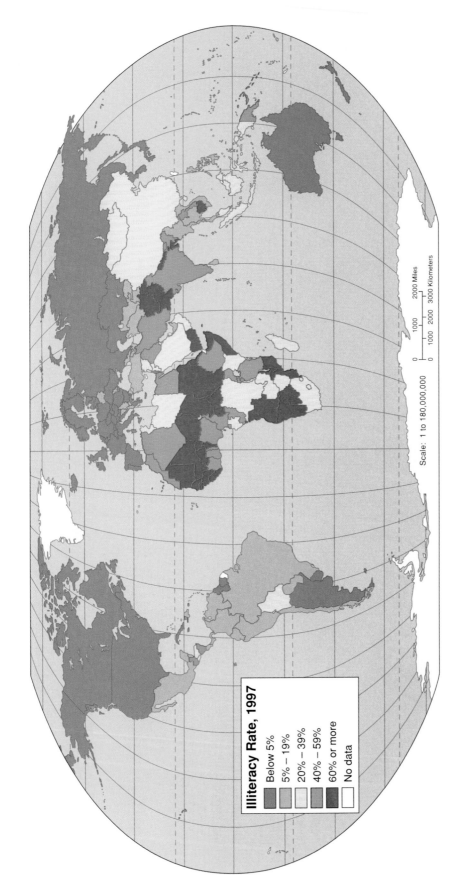

Illiteracy Rate, 1997

- Below 5%
- 5% – 19%
- 20% – 39%
- 40% – 59%
- 60% or more
- No data

Scale: 1 to 180,000,000

0   1000   2000 Miles

0   1000   2000   3000 Kilometers

The gains in living standards that developing countries have experienced during the last 2 decades are manifested in two major areas: life expectancy and literacy. The increase in global literacy is largely the consequence of an increase in primary school enrollment, particularly throughout Middle and South America, Africa, and Asia. Worldwide, education is perceived as a way to advance economic status. Unfortunately, although gains have been made, there are still countries where illiteracy rates—particularly among the females of the population—are well above global norms. The long-term potential of these countries is severely compromised as a result. **A word of caution:** Most countries view their literacy or illiteracy rates as hallmarks of their status in the world community and there is, therefore, a tendency to overstate or overestimate literacy (or, conversely, to underestimate illiteracy). In the United States for example, the stated illiteracy rate is less than 2%. Yet many experts indicate that somewhere between 10% and 15% of the U.S. population may, in fact, be functionally illiterate—that is, unable to read street signs, advertisements, or newspapers.

Map 27 Female/Male Inequality in Education and Employment

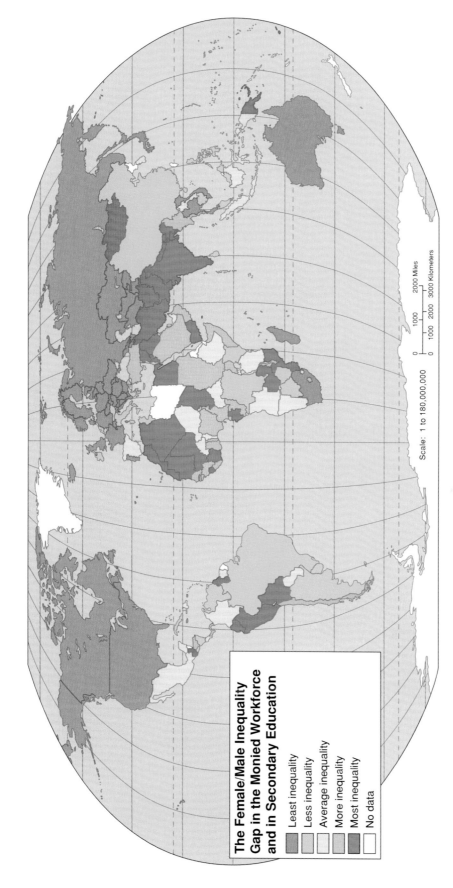

The Female/Male Inequality Gap in the Monied Workforce and in Secondary Education

- Least inequality
- Less inequality
- Average inequality
- More inequality
- Most inequality
- No data

Scale: 1 to 180,000,000

0   1000   2000 Miles
0 1000 2000 3000 Kilometers

While women in developed countries, particularly in North America and Europe, have made significant advances in socioeconomic status in recent years, in most of the world females suffer from significant inequality when compared with their male counterparts. Although women have received the right to vote in most of the world's countries, in over 90 percent of these countries that right has only been granted in the last 50 years. In most regions, literacy rates for women still fall far short of those for men; in Africa and Asia, for example, only about half as many women are as literate as men. Women marry considerably younger than men and attend school for shorter periods of time. Inequalities in education and employment are perhaps the most telling indicators of the unequal status of women in most of the world. Lack of secondary education in comparison with men prevents women from entering the workforce with equally high-paying jobs. Even where women are employed in positions similar to those held by men, they still tend to receive less compensation. The gap between rich and poor involves not only a clear geographic differentiation, but a clear gender differentiation as well.

Map 28 Quality of Life–Human Development Index

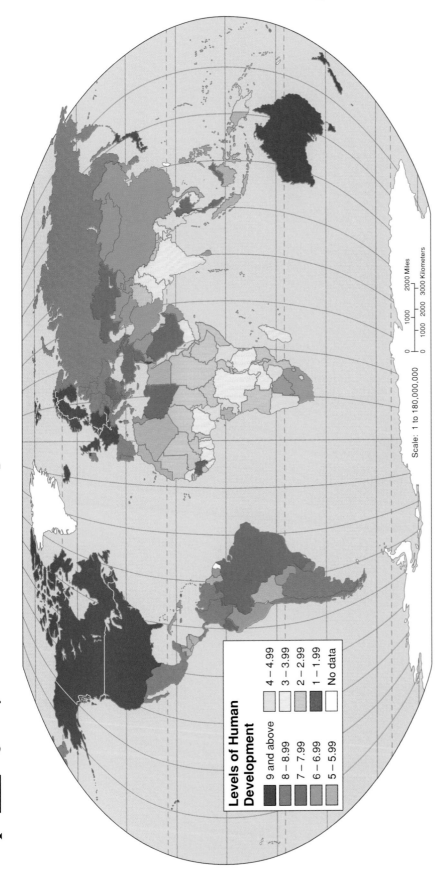

Levels of Human Development

9 and above	4 – 4.99
8 – 8.99	3 – 3.99
7 – 7.99	2 – 2.99
6 – 6.99	1 – 1.99
5 – 5.99	No data

Scale: 1 to 180,000,000

0   1000   2000 Miles
0   1000   2000   3000 Kilometers

The development index upon which this map is based takes into account a wide variety of demographic, health, and educational data, including population growth, per capita gross domestic product, longevity, literacy, and years of schooling. The map reveals significant improvement in the quality of life in Middle and South America, although it is questionable whether the gains made in those regions can be maintained in the face of the dramatic population increases expected over the next 30 years. More clearly than anything else, the map illustrates the near-desperate situation in Africa and South Asia. In those regions, the unparalleled growth in population threatens to overwhelm all efforts to improve the quality of life. In Africa, for example, the population is increasing by 20 million persons per year. With nearly 45 percent of the continent's population aged 15 years or younger, this growth rate will accelerate as the women reach childbearing age. Africa, along with South Asia, faces the very difficult challenge of providing basic access to health care, education, and jobs for a rapidly increasing population. The map also illustrates the striking difference in quality of life between those who inhabit the world's equatorial and tropical regions and those fortunate enough to live in the temperate zones, where the quality of life is significantly higher.

Part IV

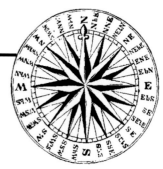

Global Economic Patterns

Map 29 Rich and Poor Countries—Gross National Product

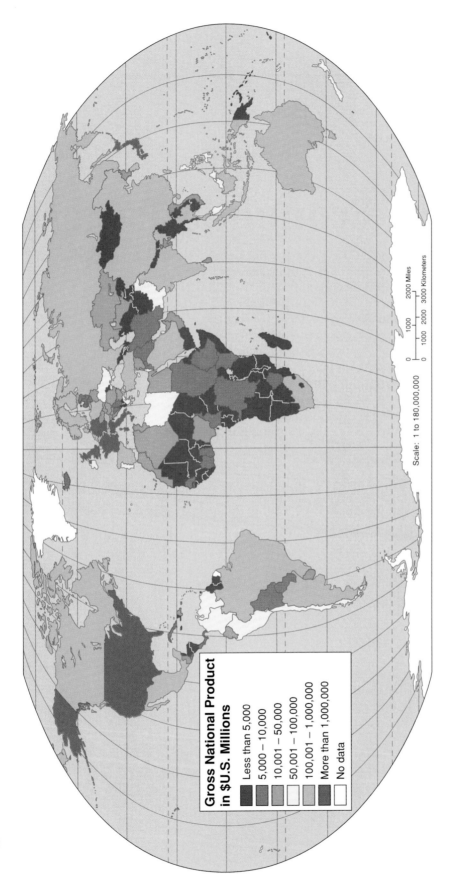

Scale: 1 to 180,000,000

0   1000   2000 Miles
0   1000   2000   3000 Kilometers

Gross National Product in $U.S. Millions

- Less than 5,000
- 5,000 – 10,000
- 10,001 – 50,000
- 50,001 – 100,000
- 100,001 – 1,000,000
- More than 1,000,000
- No data

Gross National Product (GNP) is the value of all the goods and services produced by a country, including its net income from abroad, during a year. Although GNP is often misleading and commonly incomplete, it is commonly used by economists, geographers, political scientists, policy makers, development experts, and others not only as a measure of relative well-being but as an instrument of assessing the effectiveness of economic and political policies. What is wrong with GNP? First of all, it does not take into account a number of real economic factors such as environmental deterio-ration, the accumulation or degradation of human and social capital, or the value of household work. Yet in spite of these deficiencies, GNP is still a reasonable way to assess the relative wealth of nations: the vast differences in wealth that separate the poorest countries from the richest. One of the more striking features of the map is the evidence it presents that such a small number of countries possess so many of the world's riches (keeping in mind that GNP provides no measure of the distribution of wealth within a country).

Map 30 Gross National Product Per Capita

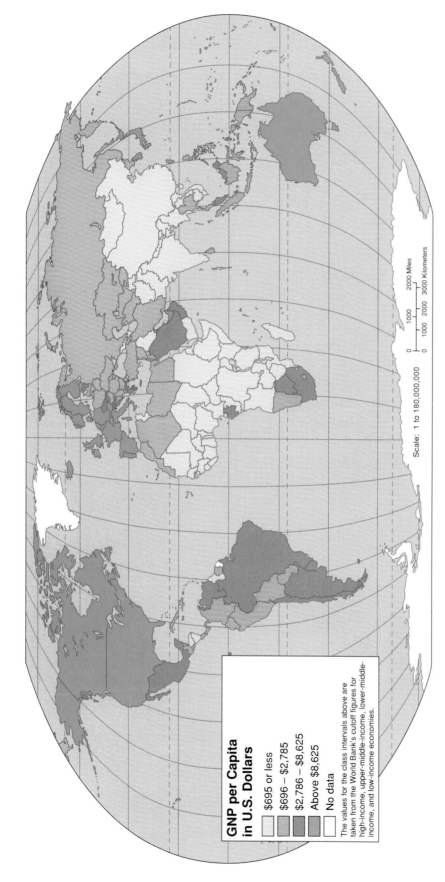

GNP per Capita in U.S. Dollars

- $695 or less
- $696 – $2,785
- $2,786 – $8,625
- Above $8,625
- No data

The values for the class intervals above are taken from the World Bank's cutoff figures for high-income, upper-middle-income, lower-middle-income, and low-income economies.

Scale: 1 to 180,000,000

0   1000   2000 Miles
0   1000   2000   3000 Kilometers

Gross National Product in either absolute or per capita form should be used cautiously as a yardstick of economic strength, because it does not measure the distribution of wealth among a population. There are countries (most notably, the oil-rich countries of the Middle East) where per capita GNP is high but where the bulk of the wealth is concentrated in the hands of a few individuals, leaving the remainder in poverty. Even within countries in which wealth is more evenly distributed (such as those in North America or western Europe), there is a tendency for dollars or pounds sterling or francs or marks to concentrate in the bank accounts of a relatively small percent of the population. Yet the maldistribution of wealth tends to be great-est in the less developed countries, where the per capita GNP is far lower than in North America and western Europe, and poverty is widespread. In fact, a map of GNP per capita offers a reasonably good picture of comparative economic well-being. It should be noted that a low per capita GNP does not automatically condemn a country to low levels of basic human needs and services. There are a few countries, such as Costa Rica and Sri Lanka, that have relatively low per capita GNP figures but rank comparatively high in other measures of human well-being, such as average life expectancy, access to medical care, and literacy.

Map 31 Economic Growth–GNP Change Per Capita

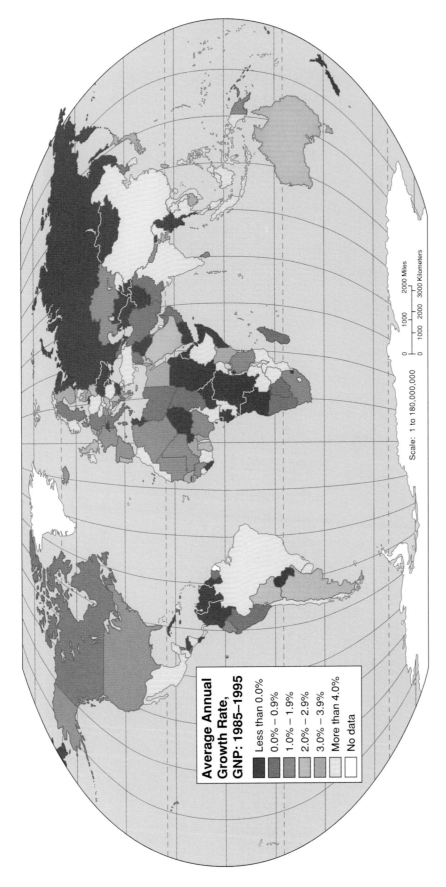

Average Annual Growth Rate, GNP: 1985–1995

- Less than 0.0%
- 0.0% – 0.9%
- 1.0% – 1.9%
- 2.0% – 2.9%
- 3.0% – 3.9%
- More than 4.0%
- No data

Scale: 1 to 180,000,000

As more and more of the world's countries move into the international trade mainstream, the share of the total global production under the control of developing countries increases. During the decade for which these data are available, developing economies in South, Southeast, and East Asia grew at rates far greater than the growth rates of the middle-income and high-income economies of Europe and North and South America. This should not necessarily be viewed as a case of the poor catching up with the rich; in fact, it shows the huge impact even relatively small production increases will have in countries with small GNPs. Nevertheless, the growth rate in the Gross National Products of some of the world's poorer countries is an encouraging trend.

Map 32 Total Labor Force

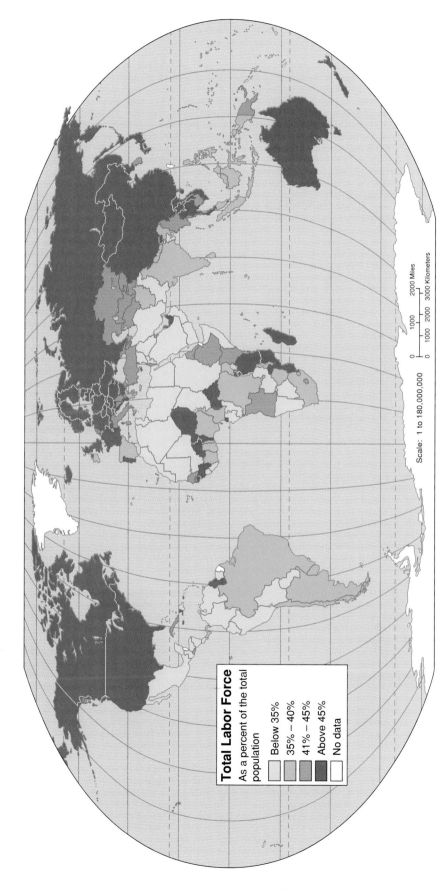

Total Labor Force

As a percent of the total population

- Below 35%
- 35% – 40%
- 41% – 45%
- Above 45%
- No data

Scale: 1 to 180,000,000

0   1000   2000 Miles

0 1000 2000 3000 Kilometers

The term *labor force* refers to the economically active portion of a population, that is, all people who work or are without work but are available for and are seeking work to produce economic goods and services. The total labor force thus includes both the employed and the unemployed (as long as they are actively seeking employment). Labor force is considered a better indicator of economic potential than employment/unemployment figures, since unemployment figures will include experienced workers with considerable potential who are temporarily out of work. Unemployment figures will also incorporate persons seeking employment for the first time (many recent college graduates, for example). Generally, countries with higher percent-

ages of total population within the labor force will be countries with higher levels of economic development. This is partly a function of levels of education and training and partly a function of the age distribution of populations. In developing countries, substantial percentages of the total population are too young to be part of the labor force. Also in developing countries a significant percentage of the population is women engaged in household activities or subsistence cultivation. These people seldom appear on lists of either employed or unemployed seeking employment and are the world's forgotten workers.

Map 33 Employment by Sector

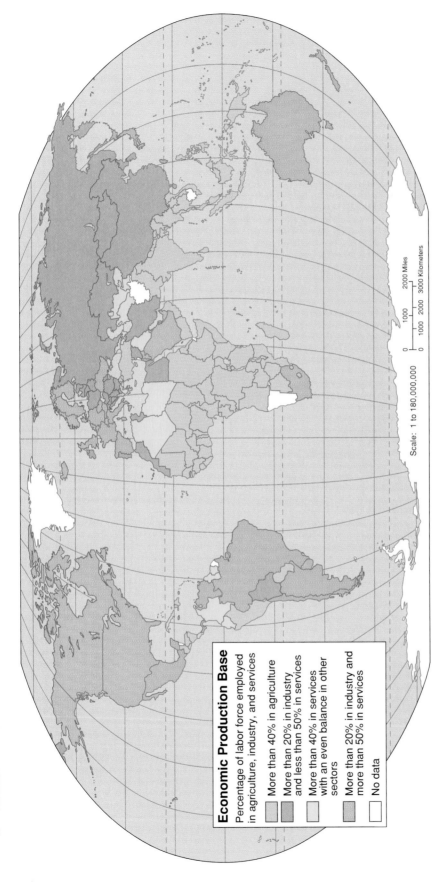

Economic Production Base

Percentage of labor force employed in agriculture, industry, and services

- More than 40% in agriculture
- More than 20% in industry and less than 50% in services
- More than 40% in services with an even balance in other sectors
- More than 20% in industry and more than 50% in services
- No data

Scale: 1 to 180,000,000

0   1000   2000 Miles

0   1000   2000   3000 Kilometers

The employment structure of a country's population is one of the best indicators of the country's position on the scale of economic development. At one end of the scale are those countries with more than 40 percent of their labor force employed in agriculture. These are almost invariably the least developed, with high population growth rates, poor human services, significant environmental problems, and so on. In the middle of the scale are two types of countries: those with more than 20 percent of their labor force employed in industry and those with a fairly even balance among agricultural, industrial, and service employment but with at least 40 percent of their labor force employed in service activities. Generally, these countries have undergone the industrial revolution fairly recently and are still devel-

oping an industrial base while building up their service activities. This category also includes countries with a disproportionate share of their economies in service activities primarily related to resource extraction. On the other end of the scale are countries from the agricultural economies are countries with more than 20 percent of their labor force employed in industry and more than 50 percent in service activities. These countries are, for the most part, those with a highly automated industrial base and a highly mechanized agricultural system (the "postindustrial," developed countries). They also include, particularly in Middle and South America and Africa, industrializing countries that are also heavily engaged in resource extraction as a service activity.

Map 34 Agricultural Production Per Capita

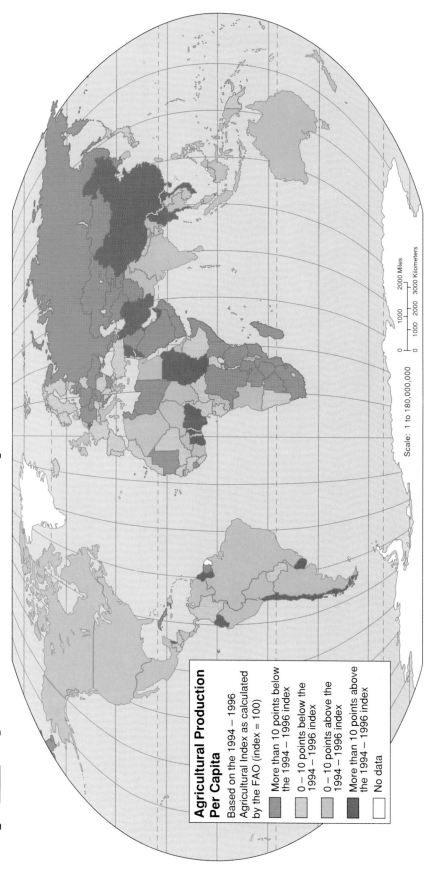

Agricultural Production Per Capita

Based on the 1994 – 1996
Agricultural Index as calculated
by the FAO (index = 100)

- More than 10 points below the 1994 – 1996 index
- 0 – 10 points below the 1994 – 1996 index
- 0 – 10 points above the 1994 – 1996 index
- More than 10 points above the 1994 – 1996 index
- No data

Scale: 1 to 180,000,000

0   1000   2000 Miles

0 1000 2000 3000 Kilometers

Agricultural production includes the value of all crop and livestock products originating within a country for the base period of 1994–1996. The index value portrays the disposable output (after deductions for livestock feed and seed for planting) of a country's agriculture in comparison with the base period 1989–1991. Thus, the production values show not only the relative ability of countries to produce food but also show whether or not that ability has increased or decreased over a 10-year period. In general, global food production has kept up with or very slightly exceeded population growth. However, there are significant regional variations in the trend of food production keeping up with or surpassing population growth. For example, agricultural production in Africa and in Middle America has fallen, while production in South America, Asia, and Europe has risen. In the case of Africa, the drop in production reflects a population growing more rapidly than agricultural productivity. Where rapid increases in food production per capita exist (as in certain countries in South America, Asia, and Europe), most often the reason is the development of new agricultural technologies that have allowed food production to grow faster than population. In much of Asia, for example, the so-called Green Revolution of new, highly productive strains of wheat and rice made positive index values possible. Also in Asia, the cessation of major warfare allowed some countries (Cambodia, Laos, and Vietnam) to show substantial increases over the 1982–1984 index. In some cases, a drop in production per capita reflects government decisions to limit production in order to maintain higher prices for agricultural products. The United States and Japan fall into this category.

Map 35 Exports of Primary Products

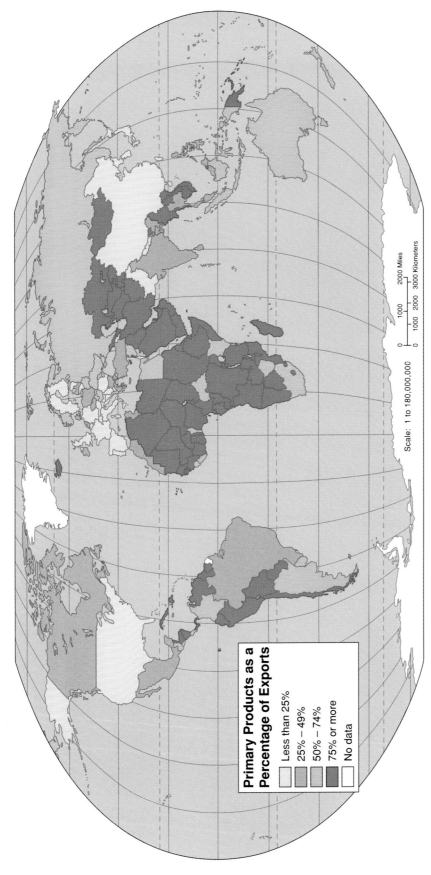

Primary Products as a Percentage of Exports

- Less than 25%
- 25% – 49%
- 50% – 74%
- 75% or more
- No data

Scale: 1 to 180,000,000

0   1000   2000 Miles

0   1000   2000   3000 Kilometers

Primary products are those that require additional processing before they enter the consumer market: metallic ores that must be converted into metals and then into metal products such as automobiles or refrigerators; forest products such as timber that must be converted to lumber before they become suitable for construction purposes; and agricultural products that require further processing before being ready for human consumption. It is an axiom in international economics that the more a country relies on primary products for its export commodities, the more vulnerable its economy is to market fluctuations. Those countries with only primary products to

export are hampered in their economic growth. A country dependent on only one or two products for export revenues is unprotected from economic shifts, particularly a changing market demand for its products. Imagine what would happen to the thriving economic status of the oil-exporting states of the Persian Gulf, for example, if an alternate source of cheap energy were found. A glance at this map shows that those countries with the lowest levels of economic development tend to be concentrated on primary products and, therefore, have economies that are especially vulnerable to economic instability.

Map 36 Dependence on Trade

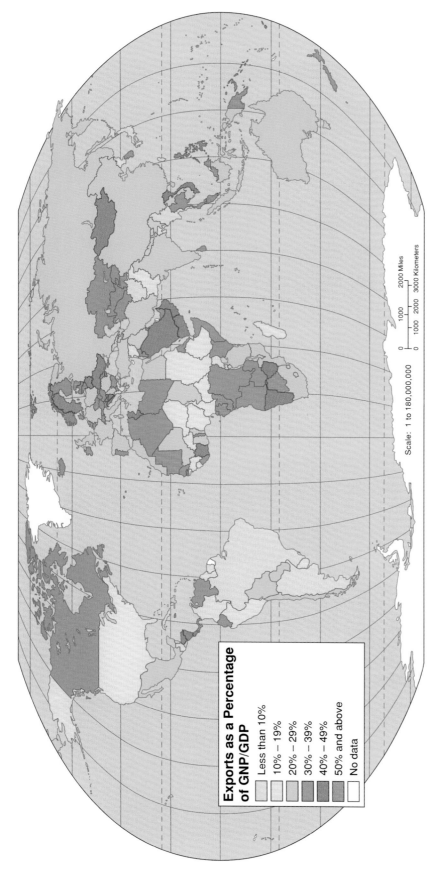

Exports as a Percentage of GNP/GDP

- Less than 10%
- 10% — 19%
- 20% — 29%
- 30% — 39%
- 40% — 49%
- 50% and above
- No data

Scale: 1 to 180,000,000

0   1000   2000 Miles

0   1000   2000   3000 Kilometers

As the global economy becomes more and more a reality, the economic strength of virtually all countries is increasingly dependent upon trade. For many developing nations, with relatively abundant resources and limited industrial capacity, exports provide the primary base upon which their economies rest. Even countries like the United States, Japan, and Germany, with huge and diverse economies, depend on exports to generate a significant percentage of their employment and wealth. Without imports, many products that consumers want would be unavailable or more expensive; without exports, many jobs would be eliminated. But exports alone do not provide the full story on trade dependence; part of what a map such as this masks is "what kind of exports?" For the more developed parts of the world, exports tend to be industrial products, perhaps along with some few raw materials (the U.S. and Russia are exceptions here in that they export a significant quantity of raw materials). But for the lesser developed countries, the exports are largely in the raw materials category. While this, as noted, provides jobs, it also means that countries may not have the necessary quantity of raw materials with which to develop their own industries and further their economic development.

Map 37 Central Government Expenditures Per Capita

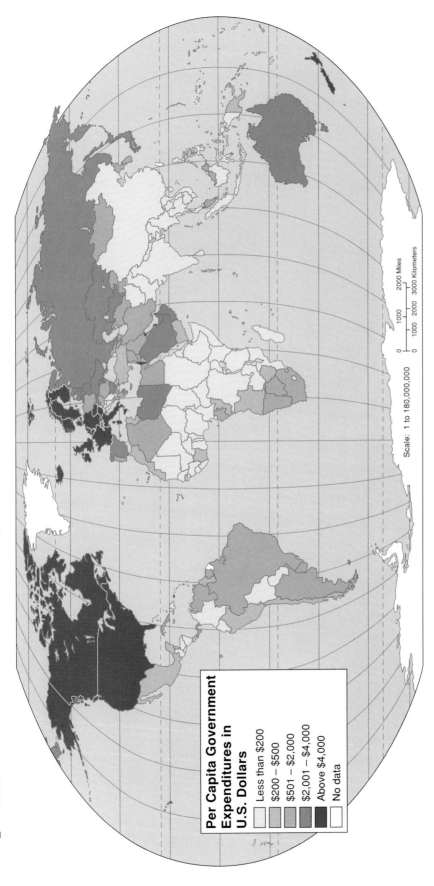

Per Capita Government Expenditures in U.S. Dollars

- Less than $200
- $200 – $500
- $501 – $2,000
- $2,001 – $4,000
- Above $4,000
- No data

Scale: 1 to 180,000,000

0 1000 2000 Miles
0 1000 2000 3000 Kilometers

The amount of money that the central government of a country spends upon a variety of essential governmental functions is a measure of relative economic development, particularly when it is viewed on a per-person basis. These functions include such governmental responsibilities as agriculture, communications, culture, defense, education, fishing and hunting, health, housing, recreation, religion, social security, transportation, and welfare. Generally, the higher the level of economic development, the greater the per capita expenditures on these services. However, the data do mask some internal variations. For example, countries that spend 20 percent or more of their central government expenditures on defense will often show up in the more developed category when, in fact, all that the figures really show is that a disproportionate amount of the money available to the government is devoted to purchasing armaments and maintaining a large standing military force. Thus, the fact that Libya spends $2,937 per capita—more than 10 times the average for Africa—does not suggest that the average Libyan is 10 times better off than the average Tanzanian. Nevertheless, this map—particularly when compared with Map 40, Energy Consumption Per Capita—does provide a reasonable approximation of economic development levels.

– 56 –

Map 38 Purchasing Power Parity

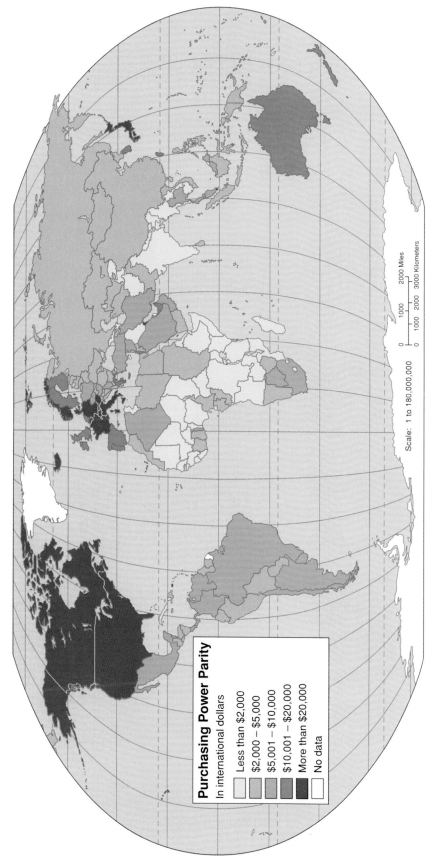

Purchasing Power Parity

In international dollars

- Less than $2,000
- $2,000 – $5,000
- $5,001 – $10,000
- $10,001 – $20,000
- More than $20,000
- No data

Scale: 1 to 180,000,000

0   1000   2000 Miles

0   1000   2000   3000 Kilometers

Among all the economic measures that separate the "haves" from the "have-nots," per capita Purchasing Power Parity (PPP) may be the most meaningful. Per capita GNP and GDP (Gross Domestic Product) figures, and even per capita income, have the limitation of seldom reflecting the true purchasing power of a country's currency at home. In order to get around this limitation, international economists seeking to compare national currencies developed the PPP measure, which shows the level of goods and services that holders of a country's money can acquire locally. By converting all currencies to the "international dollar," the World Bank and other organizations using PPP can now show more truly comparative values, since the new currency value shows the number of units of a country's currency

required to buy the same quantity of goods and services in the local market as one U.S. dollar would buy in an average country. The use of PPP currency values can alter the perceptions about a country's true comparative position in the world economy. PPP provides a valid measurement of the ability of a country's population to provide for itself the things that people in the developed world take for granted: adequate food, shelter, clothing, education, and access to medical care. A glance at the map shows a clearcut demarcation between temperate and tropical zones, with most of the countries with a PPP above $5,000 in the midlatitude zones and most of those with lower PPPs in the tropical and equatorial regions.

Map 39 Energy Production

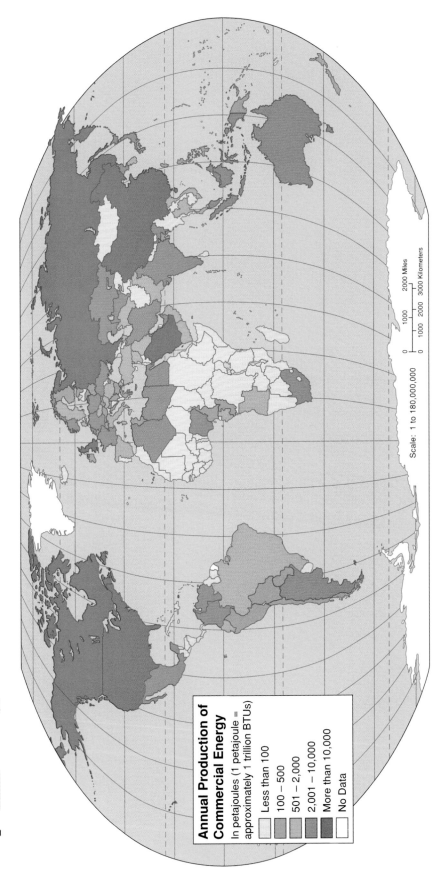

Annual Production of Commercial Energy

In petajoules (1 petajoule = approximately 1 trillion BTUs)

- Less than 100
- 100 – 500
- 501 – 2,000
- 2,001 – 10,000
- More than 10,000
- No Data

Scale: 1 to 180,000,000

0   1000   2000 Miles

0   1000   2000   3000 Kilometers

The production of commercial energy in all its forms—solid fuels (primarily coal), liquid fuels (primarily petroleum), natural gas, geothermal, wind, solar, hydroelectric, and nuclear—is a good measure of a country's ability to produce sufficient quantities of energy to meet domestic demands or to provide a healthy export commodity—or, in some instances, both. Commercial energy production is also a measure of the level of economic development, although a fairly subjective one. With exceptions, wealthier countries produce more energy from all sources than do poorer countries. Countries such as Japan and many European states rank among the world's wealthiest, but are energy-poor and produce relatively little of their own energy.

They have the ability, however, to pay for it. On the other hand, countries such as those of the Persian Gulf or the oil-producing states of Middle and South America may rank relatively low on the scale of economic development but rank high as producers of energy. The map does not show the enormous amounts of energy from noncommercial sources (traditional fuels like firewood and animal dung) used by the world's poor, particularly in Middle and South America, Africa, South Asia, and East Asia. In these regions, firewood and animal dung may account for more actual energy production than coal or oil. Indeed, for many in the developing world, the real energy crisis is a shortage of wood for cooking and heating.

Map 40 Energy Consumption Per Capita

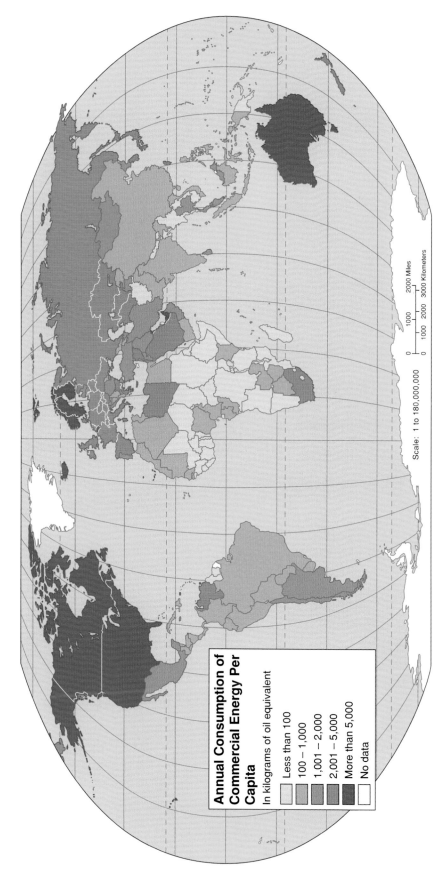

Annual Consumption of Commercial Energy Per Capita

In kilograms of oil equivalent

- Less than 100
- 100 – 1,000
- 1,001 – 2,000
- 2,001 – 5,000
- More than 5,000
- No data

Scale: 1 to 180,000,000

0   1000   2000 Miles

0 1000 2000 3000 Kilometers

Of all the quantitative measures of economic well-being, energy consumption per capita may be the most expressive. All of the countries defined by the World Bank as having high incomes consume at least 100 gigajoules of commercial energy (the equivalent of about 3.5 metric tons of coal) per person per year, with some, such as the United States and Canada, having consumption rates in the 300 gigajoule range (the equivalent of more than 10 metric tons of coal per person per year). With the exception of the oil-rich Persian Gulf states, where consumption figures include the costly "burning off" of excess energy in the form of natural gas flares at wellheads, most of the highest-consuming countries are in the Northern Hemisphere, concentrated in North America and Western Europe. At the other end of the scale are low-income countries, whose consumption rates are often less than 1 percent of those of the United States and other high consumers. These figures do not, of course, include the consumption of noncommercial energy—the traditional fuels of firewood, animal dung, and other organic matter—widely used in the less developed parts of the world.

Part V

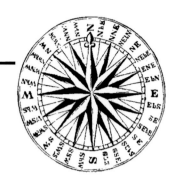

Political Systems

Map 41 Political Systems

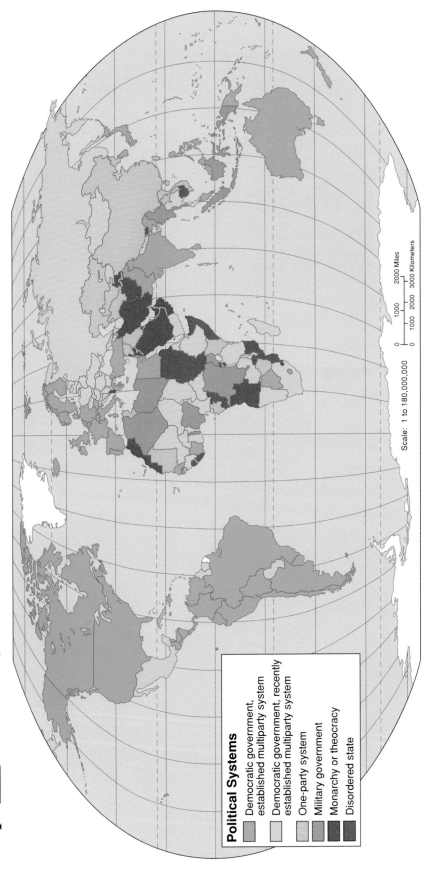

Scale: 1 to 180,000,000

Political Systems	
	Democratic government, established multiparty system
	Democratic government, recently established multiparty system
	One-party system
	Military government
	Monarchy or theocracy
	Disordered state

The world map of political systems has changed dramatically during the 1990s. The categories of political systems shown are subject to some interpretation: established multiparty democracies are those in which elections by secret ballot with adult suffrage are long-term features of the political landscape; recently established multiparty democracies are those in which this distinguishing feature has only recently emerged. The former Soviet satellites of eastern Europe and the republics that formerly constituted the USSR are in this category; so are states in emerging regions that are beginning to throw off the single-party rule that often followed the violent upheavals of the immediate postcolonial governmental transitions. The other categories are more-or-less obvious. One-party systems define states where single party rule provided for by constitution is a fact of political life.

Monarchies are countries with heads of state who are members of a royal family. Some countries with monarchs do not fall into this category because their monarchs are titular heads of state only. Theocracies are countries in which rule is within the hands of a priestly or clerical class; for example, fundamentalist Islamic countries. Military governments are often organized around a junta that has seized control of the government from civil authority; such states are often technically transitional—the military claims that it will return the reins of government to civil authority "when order is restored." Finally, "disordered states" are countries so beset by civil war or widespread ethnic conflict that no organized government can be said to exist within them.

– 62 –

Map 42 Sovereign States 1998

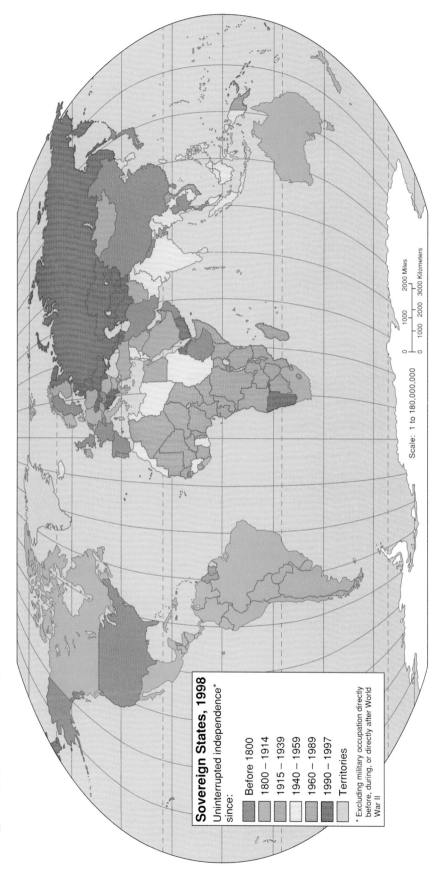

Sovereign States, 1998

Uninterrupted independence* since:

- Before 1800
- 1800 – 1914
- 1915 – 1939
- 1940 – 1959
- 1960 – 1989
- 1990 – 1997
- Territories

* Excluding military occupation directly before, during, or directly after World War II

Scale: 1 to 180,000,000

0   1000   2000 Miles
0   1000   2000   3000 Kilometers

Most countries of the modern world, including such major states as Germany and Italy, became independent after the beginning of the nineteenth century. Of the world's current countries, only 27 were independent in 1800. (Ten of the 27 were in Europe; the others were Afghanistan, China, Colombia, Ethiopia, Haiti, Iran, Japan, Mexico, Nepal, Oman, Paraguay, Russia, Taiwan, Thailand, Turkey, the United States, and Venezuela.) Following 1800, there have been four great periods of national independence. During the first of these (1800–1914), most of the mainland countries of the Americas achieved independence. During the second period (1915–1939), the countries of Eastern Europe emerged as independent entities. The third period (1940–1959) includes World War II and the years that followed, when independence for African and Asian nations that had been under control of colonial powers first began to occur. During the fourth period (1960–1989), independence came to the remainder of the colonial African and Asian nations, as well as to former colonies in the Caribbean and the South Pacific. More than half of the world's countries came into being as independent political entities during this period. Finally, in the last few years (1990–1997), the breakup of the existing states of the Soviet Union, Yugoslavia, and Czechoslovakia created 22 countries where only 3 had existed before.

Map 43 Post-Cold War International Alliances

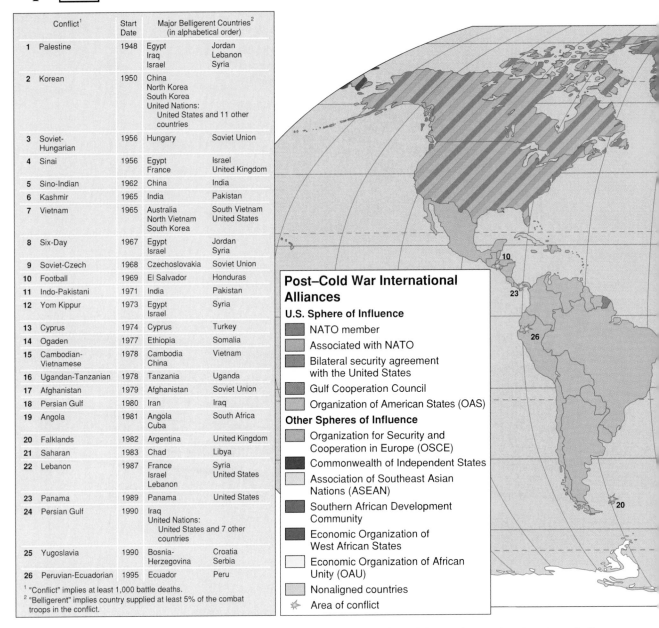

	Conflict[1]	Start Date	Major Belligerent Countries[2] (in alphabetical order)	
1	Palestine	1948	Egypt Iraq Israel	Jordan Lebanon Syria

2	Korean	1950	China
			North Korea
			South Korea
			United Nations:
			  United States and 11 other
			  countries	

3	Soviet-Hungarian	1956	Hungary	Soviet Union
4	Sinai	1956	Egypt France	Israel United Kingdom
5	Sino-Indian	1962	China	India
6	Kashmir	1965	India	Pakistan
7	Vietnam	1965	Australia North Vietnam South Korea	South Vietnam United States
8	Six-Day	1967	Egypt Israel	Jordan Syria
9	Soviet-Czech	1968	Czechoslovakia	Soviet Union
10	Football	1969	El Salvador	Honduras
11	Indo-Pakistani	1971	India	Pakistan
12	Yom Kippur	1973	Egypt Israel	Syria
13	Cyprus	1974	Cyprus	Turkey
14	Ogaden	1977	Ethiopia	Somalia
15	Cambodian-Vietnamese	1978	Cambodia China	Vietnam
16	Ugandan-Tanzanian	1978	Tanzania	Uganda
17	Afghanistan	1979	Afghanistan	Soviet Union
18	Persian Gulf	1980	Iran	Iraq
19	Angola	1981	Angola Cuba	South Africa
20	Falklands	1982	Argentina	United Kingdom
21	Saharan	1983	Chad	Libya
22	Lebanon	1987	France Israel Lebanon	Syria United States
23	Panama	1989	Panama	United States

24	Persian Gulf	1990	Iraq
			United Nations:
			  United States and 7 other
			  countries	

25	Yugoslavia	1990	Bosnia-Herzegovina	Croatia Serbia
26	Peruvian-Ecuadorian	1995	Ecuador	Peru

[1] "Conflict" implies at least 1,000 battle deaths.
[2] "Belligerent" implies country supplied at least 5% of the combat troops in the conflict.

Post–Cold War International Alliances

U.S. Sphere of Influence
- NATO member
- Associated with NATO
- Bilateral security agreement with the United States
- Gulf Cooperation Council
- Organization of American States (OAS)

Other Spheres of Influence
- Organization for Security and Cooperation in Europe (OSCE)
- Commonwealth of Independent States
- Association of Southeast Asian Nations (ASEAN)
- Southern African Development Community
- Economic Organization of West African States
- Economic Organization of African Unity (OAU)
- Nonaligned countries
- ✵ Area of conflict

Following World War II there developed an elaborate international system of alliances based on potential or perceived military threats. The most important of these were NATO (the North Atlantic Treaty Organization) and the Warsaw Pact (the Soviet Union and its eastern European satellite nations). The Warsaw Pact ceased to exist in 1992, leaving NATO as the only major military alliance in the world. Indeed, a number of former Soviet satellite nations are in the process of becoming NATO pact members and there is some speculation that Russia herself may become a NATO member at some point in the future. During the years that NATO and the Warsaw Pact dominated global military policy making, there were numerous conflicts involving military forces of belligerent countries. Most of these conflicts inflicted heavy damage to the environment and to civilian populations, as well as casualties among the participating military forces. Many of these conflicts were not wars in the traditional sense, in which two or more countries formally declare war upon one another, severing diplomatic ties and devoting their entire national energies to the war effort (as most belligerent nations did in World War II). Rather, many of these conflicts

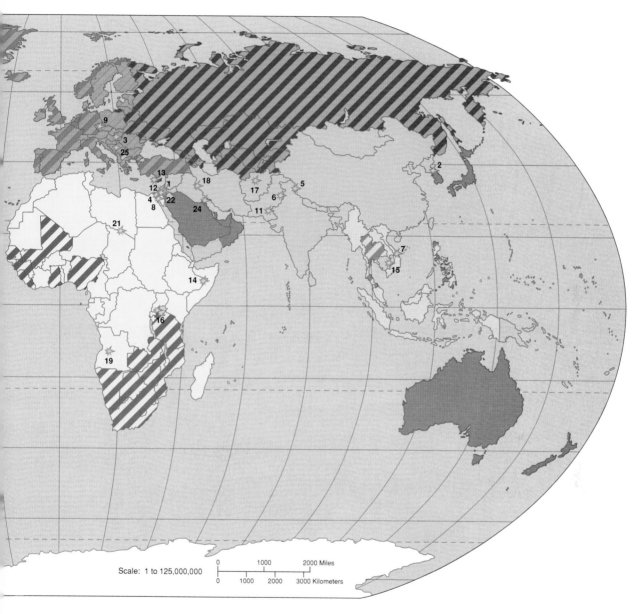

were undeclared wars, often fought between rival groups within the same country with outside support from other countries. These conflicts included civil wars, insurgencies, or wars produced by irredentism or separatist groups within a country. The Afghan war, for example, was fought between the formal government of Afghanistan and rebellious or insurgency forces, with the Soviet Union coming to the aid of the formal government. Similarly, the Vietnam war was really a civil war between rival Vietnamese factions for control of Vietnam, with the United States entering the conflict in support of the southern Vietnamese. The conflict in the fragmented remnants of Yugoslavia is another example of irredentism, although in this instance the conflict spread throughout a number of countries that had formally become independent states. Other alliances with at least some military implications include ANZUS (Australia, New Zealand, the United States), OAS (the Organization of American States), OAU (the Organization of African Unity), and OSCE (the Organization for Security and Cooperation in Europe).

Map 44 Global Distribution of Minority Groups

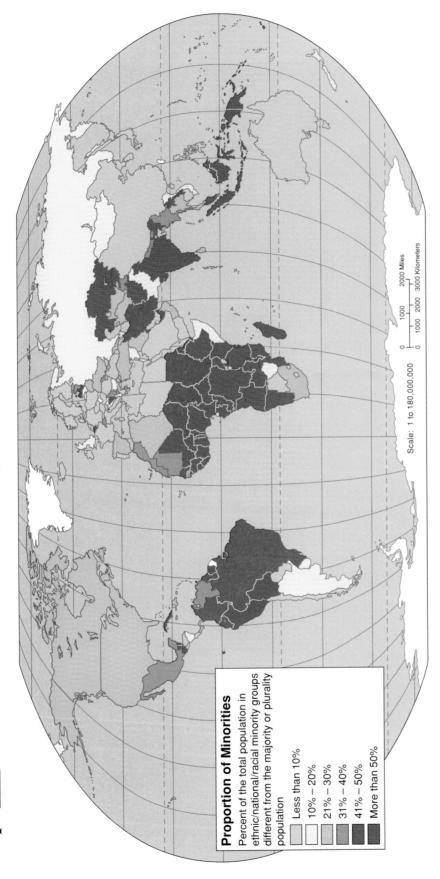

Proportion of Minorities

Percent of the total population in ethnic/national/racial minority groups different from the majority or plurality population

Less than 10%
10% – 20%
21% – 30%
31% – 40%
41% – 50%
More than 50%

Scale: 1 to 180,000,000

0   1000   2000 Miles
0   1000   2000   3000 Kilometers

The presence of minority ethnic, national, or racial groups within a country's population can add a vibrant and dynamic mix to the whole. Plural societies with a high degree of cultural and ethnic diversity should, according to some social theorists, be among the world's most healthy. Unfortunately, the reality of the situation is quite different from theory or expectation. The presence of significant minority populations played an important role in the disintegration of the Soviet Union; the continuing existence of minority populations within the new states formed from former Soviet republics threatens the viability and stability of those young political units. In Africa, national boundaries were drawn by colonial powers without regard for the geographi-

cal distribution of ethnic groups, and the continuing tribal conflicts that have resulted hamper both economic and political development. Even in the most highly developed regions of the world, the presence of minority ethnic populations poses significant problems: witness the separatist movement in Canada, driven by the desire of some French-Canadians to be independent of the English majority, and the continuing ethnic conflict between Flemish-speaking and Walloon-speaking Belgians. This map, by arraying states on a scale of homogeneity to heterogeneity, indicates areas of existing and potential social and political strife.

Map 45 World Refugee Population

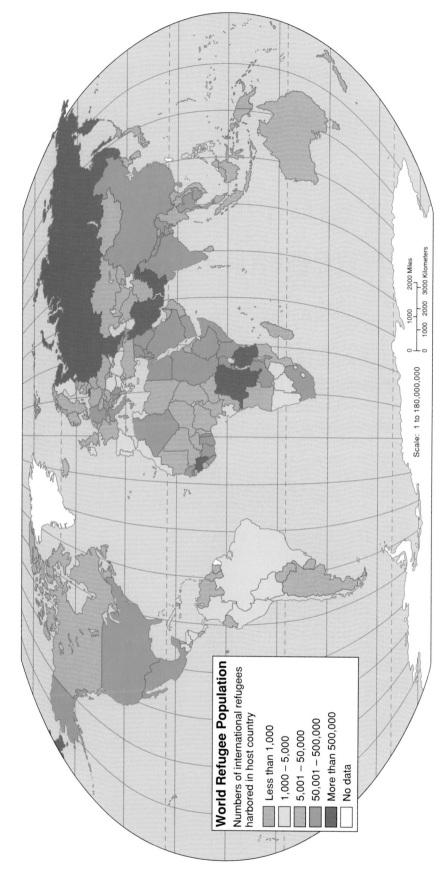

World Refugee Population

Numbers of international refugees
harbored in host country

- Less than 1,000
- 1,000 – 5,000
- 5,001 – 50,000
- 50,001 – 500,000
- More than 500,000
- No data

Scale: 1 to 180,000,000

0   1000   2000 Miles
0   1000   2000   3000 Kilometers

Refugees are persons who have been driven from their homes, normally by armed conflict, and have sought refuge by relocating. The most numerous refugees have traditionally been international refugees, who have crossed the political boundaries of their homelands into other countries. This refugee population is recognized by international agencies, and the countries of refuge are often rewarded financially by those agencies for their willingness to take in externally displaced persons. In recent years, largely because of an increase in civil wars, there have been growing numbers of internally displaced persons—those who leave their homes but stay within their country of origin. There are no rewards for harboring such internal refugee populations.

Part VI

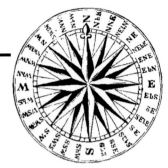

Global Patterns of Environmental Disturbance

Map 46 Global Air Pollution: Sources and Wind Currents

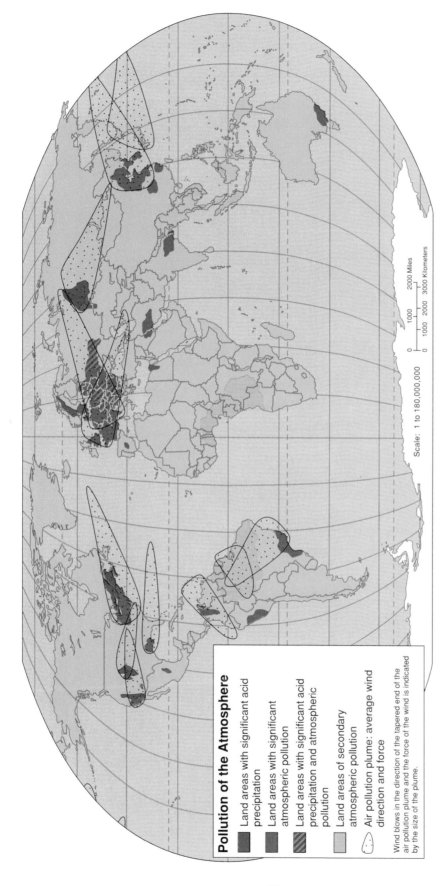

Pollution of the Atmosphere

Land areas with significant acid precipitation

Land areas with significant atmospheric pollution

Land areas with significant acid precipitation and atmospheric pollution

Land areas of secondary atmospheric pollution

Air pollution plume: average wind direction and force

Wind blows in the direction of the tapered end of the air pollution plume and the force of the wind is indicated by the size of the plume.

Scale: 1 to 180,000,000

0   1000   2000 Miles

0   1000   2000   3000 Kilometers

Almost all processes of physical geography begin and end with the flows of energy and matter among land, sea, and air. Because of the primacy of the atmosphere in this exchange system, air pollution is potentially one of the most dangerous human modifications in environmental systems. Pollutants such as various oxides of nitrogen or sulfur cause the development of acid precipitation, that damages soil, vegetation, and wildlife and fish. Air pollu-

tion in the form of smog is often dangerous for human health. And most atmospheric scientists believe that the efficiency of the atmosphere in retaining heat—the so-called "greenhouse effect"—is being enhanced by increased carbon dioxide, methane, and other gases produced by agricultural and industrial activities. The result, they fear, will be a period of global warming that will dramatically alter climates in all parts of the world.

Map 47 The Acid Deposition Problem: Air, Water, Soil

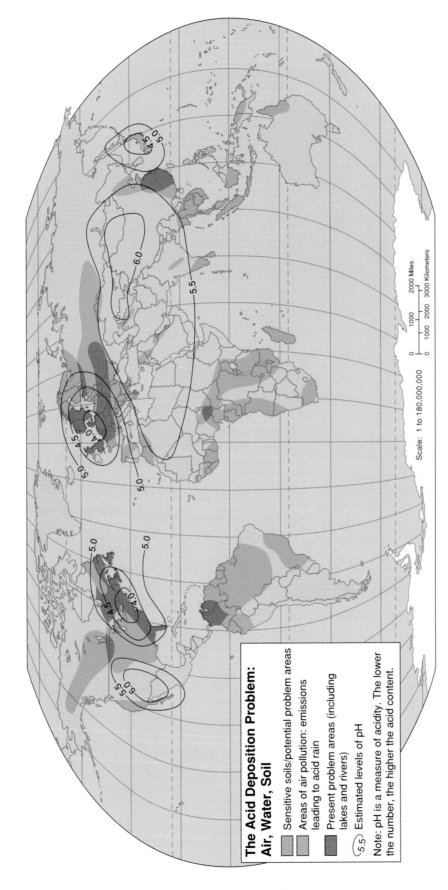

The Acid Deposition Problem: Air, Water, Soil

Sensitive soils/potential problem areas

Areas of air pollution: emissions leading to acid rain

Present problem areas (including lakes and rivers)

5.5 Estimated levels of pH

Note: pH is a measure of acidity. The lower the number, the higher the acid content.

Scale: 1 to 180,000,000

0   1000   2000 Miles
0   1000 2000 3000 Kilometers

The term "acid precipitation" refers to increasing levels of acidity in snowfall and rainfall caused by atmospheric pollution. Oxides of nitrogen and sulfur resulting from incomplete combustion of fossil fuels (coal, oil, and natural gas) combine with water vapor in the atmosphere to produce weak acids that then "precipitate" or fall along with water or ice crystals. Some atmospheric acids formed by this process are known as "dry-acid" precipitates and they too will fall to earth, although not necessarily along with rain or snow. In some areas of the world, the increased acidity of streams and lakes stemming from high levels of acid precipitation or dry acid fallout has damaged or destroyed aquatic life. Acid precipitation and dry acid fallout

also harms soil systems and vegetation, producing a characteristic burned appearance in forests that lends the same quality to landscapes that forest fires would. The region most dramatically impacted by acid precipitation is Central Europe where decades of destructive environmental practices, including the burning of high sulfur coal for commercial, industrial, and residential purposes has produced the destruction of hundreds of thousands of acres of woodlands—a phenomenon described by the German foresters who began their study of the area following the lifting of the Iron Curtain as "Waldsterben": Forest Death.

Map 48 Major Polluters and Common Pollutants

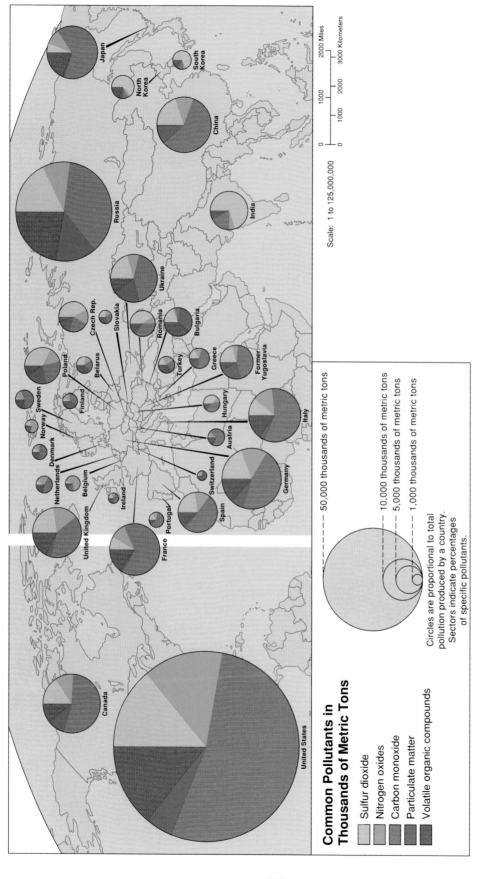

Common Pollutants in Thousands of Metric Tons

- Sulfur dioxide
- Nitrogen oxides
- Carbon monoxide
- Particulate matter
- Volatile organic compounds

Circles are proportional to total pollution produced by a country. Sectors indicate percentages of specific pollutants.

- 50,000 thousands of metric tons
- 10,000 thousands of metric tons
- 5,000 thousands of metric tons
- 1,000 thousands of metric tons

Scale: 1 to 125,000,000

0 1000 2000 3000 Kilometers

0 1000 2000 Miles

More than 90 percent of the world's total of anthropogenic (human-generated) air pollutants come from the heavily populated industrial regions of North America, Europe, South Asia (primarily in India) and East Asia (mainly in China, Japan, and the two Koreas). This map shows the origins of the five most common pollutants: sulfur dioxide, nitrogen oxide, carbon monoxide, particulate matter, and volatile organic compounds. These substances are produced both by industry and by the combustion of fossil fuels that generate electricity and power trains, planes, automobiles, buses, and trucks. In addition to combining with other components of the atmosphere and with one another to produce smog, they are the chief ingredients in acid accumulations in the atmosphere, which ultimately result in acid deposition, either as acid precipitation or dry acid fallout. Like other forms of pollutants, these air pollutants do not recognize political boundaries, and regions downwind of major polluters receive large quantities of pollutants from areas over which they often have no control.

Map 49 Global Carbon Dioxide Emissions

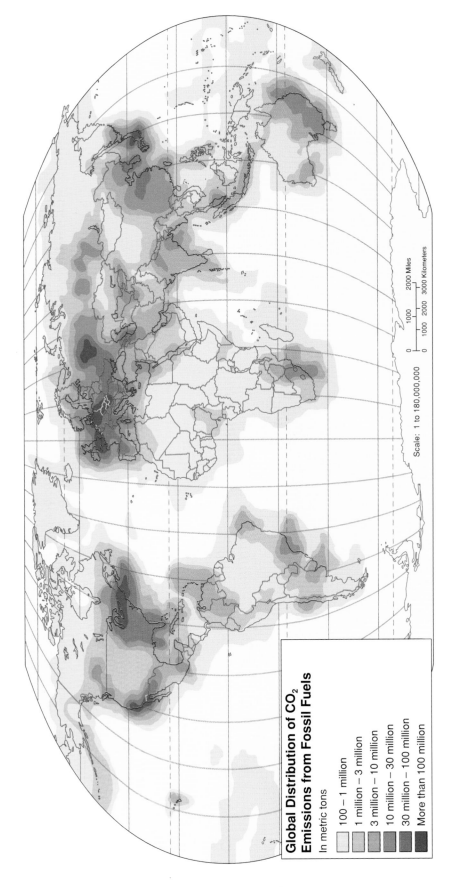

Global Distribution of CO₂ Emissions from Fossil Fuels

In metric tons

- 100 − 1 million
- 1 million − 3 million
- 3 million − 10 million
- 10 million − 30 million
- 30 million − 100 million
- More than 100 million

Scale: 1 to 180,000,000

0   1000   2000 Miles
0   1000   2000   3000 Kilometers

One of the most important components of the atmosphere is the gas carbon dioxide, the byproduct of animal respiration, of decomposition, and of combustion. During the past 200 years, atmospheric CO₂ has risen dramatically, largely as the result of the tremendous increase in fossil fuel combustion brought on by the industrialization of the world's economy and the burning and clearing of forests by the expansion of farming. While carbon dioxide by itself is relatively harmless, it is an important "greenhouse gas." The gases in the atmosphere act like the panes of glass in a greenhouse roof, allowing light in but preventing heat from escaping. The greenhouse capacity of the atmosphere is crucial for organic life and is a purely natural component of the global energy cycle. But too much carbon dioxide and other greenhouse gases such as methane could cause the earth's atmosphere to warm up too much, producing the global warming that atmospheric scientists are concerned about. Researchers estimate that if greenhouse gases such as carbon dioxide continue to increase at their present rates, the earth's mean temperature could rise between 1.5 and 4.5 degrees Celsius by the middle of the next century. Such a rise in global temperatures would produce massive alterations in the world's climate patterns.

− 72 −

Map 50 Potential Global Temperature Change

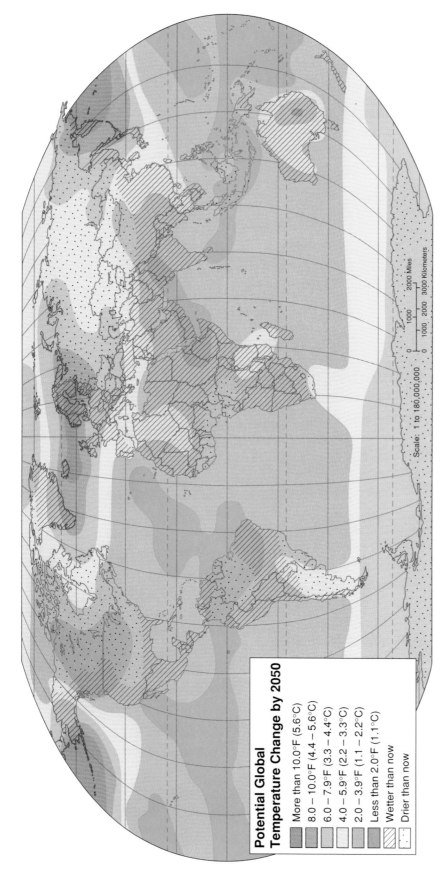

Potential Global Temperature Change by 2050

- More than 10.0°F (5.6°C)
- 8.0 – 10.0°F (4.4 – 5.6°C)
- 6.0 – 7.9°F (3.3 – 4.4°C)
- 4.0 – 5.9°F (2.2 – 3.3°C)
- 2.0 – 3.9°F (1.1 – 2.2°C)
- Less than 2.0°F (1.1°C)
- Wetter than now
- Drier than now

Scale: 1 to 180,000,000

0   1000   2000 Miles

0   1000   2000   3000 Kilometers

According to atmospheric scientists, one of the major problems of the twenty-first century will be "global warming," produced as the atmosphere's natural ability to trap and retain heat is enhanced by increased percentages of carbon dioxide, methane, chlorinated fluorocarbons or "CFCs," and other "greenhouse gases" in the earth's atmosphere. Computer models based on atmospheric percentages of carbon dioxide resulting from present use of fossil fuels show that warming is not just a possibility but a probability. Increased temperatures would cause precipitation patterns to alter significantly as well and would produce a number of other harmful effects, including a rise in the level of the world's oceans that could flood most coastal cities. International conferences on the topic of the enhanced greenhouse effect, have resulted in several international agreements to reduce the emission of carbon dioxide or to maintain it at present levels. Unfortunately, the solution is not that simple since reduction of carbon dioxide emissions is, in the short run, expensive—particularly as long as the world's energy systems continue to be based on fossil fuels. Chief among the countries that could be hit by serious international mandates to reduce emissions are those highest on the development scale who use the highest levels of fossil fuels and, therefore, produce the highest emissions, and those on the lowest end of the development scale whose efforts to industrialize could be severely impeded by the more expensive energy systems that would replace fossil fuels.

Map 51 World Water Resources: Availability of Renewable Water Per Capita

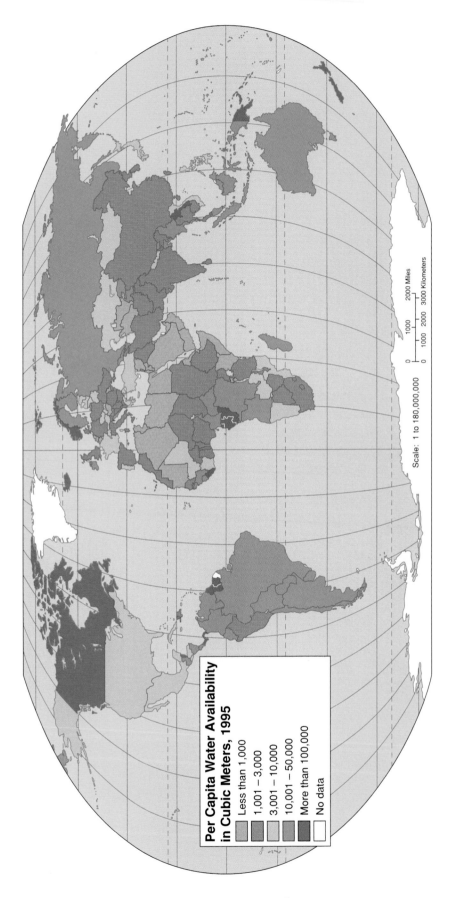

Per Capita Water Availability in Cubic Meters, 1995

- Less than 1,000
- 1,001 – 3,000
- 3,001 – 10,000
- 10,001 – 50,000
- More than 100,000
- No data

Scale: 1 to 180,000,000

0   1000   2000 Miles
0 1000 2000 3000 Kilometers

Renewable water resources are usually defined as the total water available from streams and rivers (including flows from other countries), ponds and lakes, and groundwater storage or aquifers. Not included in the total of renewable water would be water that comes from such nonrenewable sources as desalinization plants or melted icebergs. While the concept of renewable or flow resources is a traditional one in resource management, in fact, few resources, including water, are truly renewable when their use is excessive. The water resources shown here are indications of that principle. A country like the United States possesses truly enormous quantities of water. But the United States also uses enormous quantities of water. The result is availability of renewable water that is—largely because of excess use—much less than in many other parts of the world where the total supply of water is significantly less.

Map 52 World Water Resources: Annual Withdrawal Per Capita

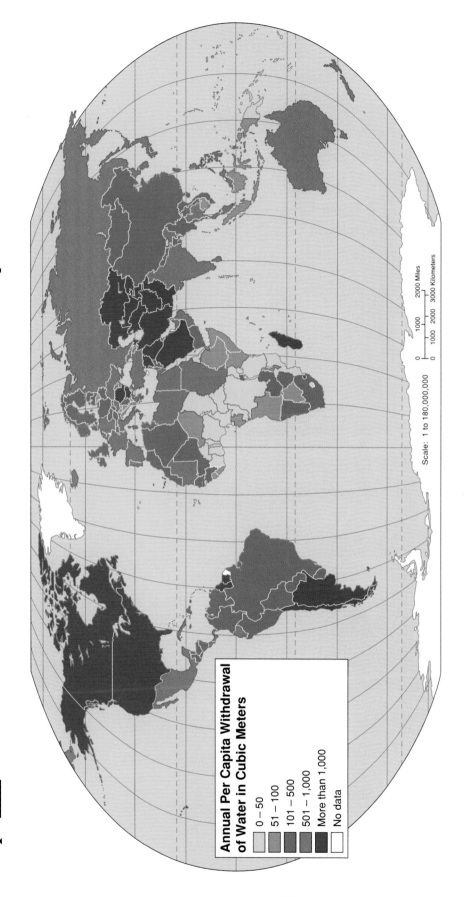

Annual Per Capita Withdrawal of Water in Cubic Meters

- 0 – 50
- 51 – 100
- 101 – 500
- 501 – 1,000
- More than 1,000
- No data

Scale: 1 to 180,000,000

0   1000   2000 Miles
0   1000   2000   3000 Kilometers

Water resources must be viewed like a bank account in which deposits and withdrawals are made. As long as the deposits are greater than the withdrawals, a positive balance remains. But when the withdrawals begin to exceed the deposits, sooner or later (depending on the relative sizes of the deposits and withdrawals) the account becomes overdrawn. For many of the world's countries, annual availability of water is insufficient to cover the demand. In these countries, reserves stored in groundwater are being tapped, resulting in depletion of the water supply (think of this as shifting money from a savings account to a checking account). The water supply can maintain its status as a renewable resource only if deposits continue to be greater than withdrawals, and that seldom happens. In general, countries with high levels of economic development and countries that rely on irrigation agriculture are the most spendthrift when it comes to their water supplies.

Map 53 Pollution of the Oceans

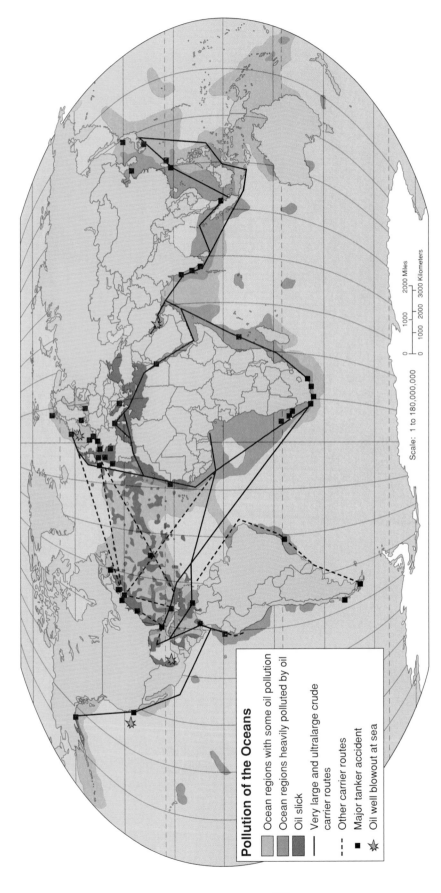

Pollution of the Oceans

	Ocean regions with some oil pollution
	Ocean regions heavily polluted by oil
	Oil slick
———	Very large and ultralarge crude carrier routes
———	Other carrier routes
- - -	Major tanker accident
■	Oil well blowout at sea
✦	

Scale: 1 to 180,000,000

0   1000   2000 Miles

0   1000   2000   3000 Kilometers

The pollution of the world's oceans has long been a matter of concern to physical geographers, oceanographers, and other environmental scientists. The great circulation systems of the ocean (shown in Map 3a) are one of the controlling factors of the earth's natural environment, and modifications to those systems have unknown consequences. This map is based on what we can measure: (1) areas of oceans where oil pollution has been proven to have inflicted significant damage to ocean ecosystems and life-forms (including phytoplankton, the oceans' primary food producers, equivalent to land-based vegetation) and (2) areas of oceans where unusually high con-

centrations of hydrocarbons from oil spills may have inflicted some damage to the oceans' biota. A glance at the map shows that there are few areas of the world's oceans where some form of pollution is not a part of the environmental system. What the map does not show in detail, because of the scale, are the dramatic consequences of large individual pollution events: the wreck of the *Exxon Valdez* and the polluting of Prince William Sound or the environmental devastation produced by the Gulf War in the Persian Gulf.

– 76 –

Map 54 Food Supply from the World's Marine and Freshwater Systems

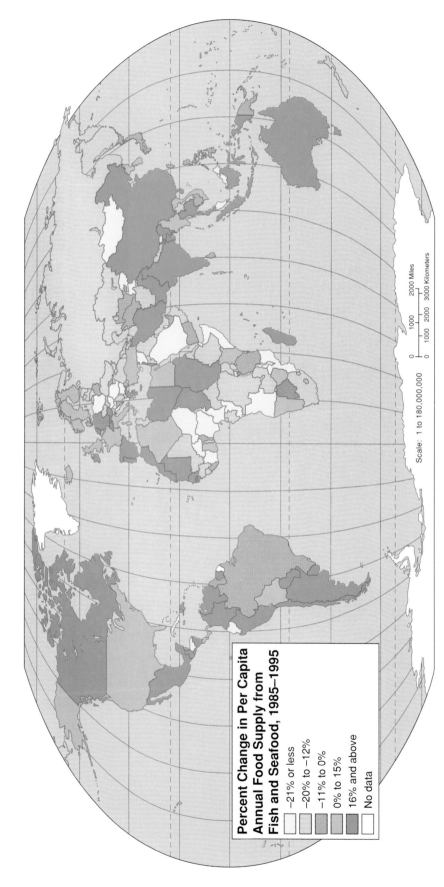

Percent Change in Per Capita Annual Food Supply from Fish and Seafood, 1985–1995

- –21% or less
- –20% to –12%
- –11% to 0%
- 0% to 15%
- 16% and above
- No data

Scale: 1 to 180,000,000

0   1000   2000 Miles

0   1000   2000   3000 Kilometers

Not that many years ago, food supply experts were confidently predicting that the "starving millions" of the world of the future could be fed from the unending bounty of the world's oceans. While the annual catch from the sea helped to keep hunger at bay for a time, by the late 1980s it had become apparent that without serious human intervention in the form of aquaculture, the supply of fish would not be sufficient to offset the population/food imbalance that was beginning to affect so many of the world's regions. The development of factory-fishing with advanced equipment to locate fish and process them before they went to market increased the supply of food from the ocean, but in that increase was sown the seeds of future problems. The factory-fishing system, efficient in terms of econom-

ics, was costly in terms of fish populations. In some well-fished areas, the stock of fish that was viewed as near infinite just a few decades ago has dwindled nearly to the point of disappearance. This map shows both increases and decreases in the amount of individual countries' food supplies from the ocean. The increases are often the result of more technologically advanced fishing operations. The decreases are usually the result of the same thing: increased technology has brought increased harvests which has reduced the supply of fish and shellfish and that, in turn, has increased prices. Most of the countries that have experienced sharp decreases in their supply of food from the world's oceans are simply no longer able to pay for an increasingly scarce commodity.

Map 55 Cropland Per Capita: Changes, 1985–1995

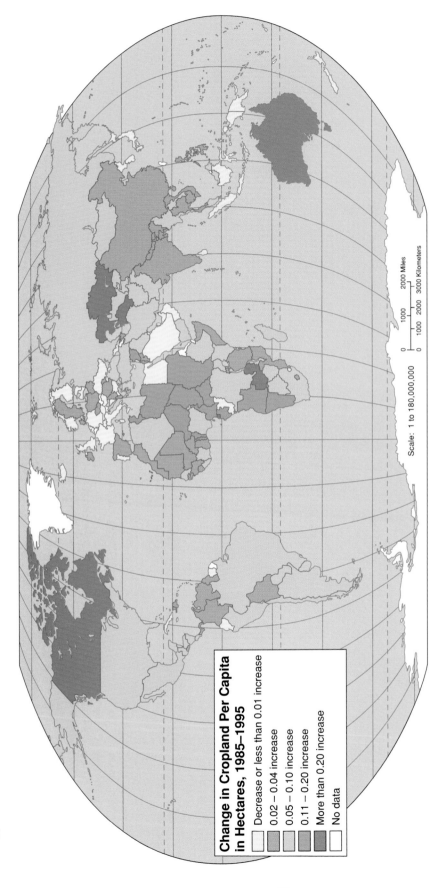

**Change in Cropland Per Capita
in Hectares, 1985–1995**

- Decrease or less than 0.01 increase
- 0.02 – 0.04 increase
- 0.05 – 0.10 increase
- 0.11 – 0.20 increase
- More than 0.20 increase
- No data

Scale: 1 to 180,000,000

0   1000   2000 Miles

0   1000   2000   3000 Kilometers

As population has increased rapidly throughout the world, area in cultivated land has increased at the same time; in fact, the amount of farmland per person has gone up slightly. Unfortunately, the figures that show this also tell us that since most of the best (or even good) agricultural land in 1985 was already under cultivation, most of the agricultural area added since the early 1980s involves land that would have been viewed as marginal by the fathers and grandfathers of present farmers—marginal in that it was too dry, too wet, too steep to cultivate, too far from a market, and so on.

The continued expansion of agricultural area is one reason that serious famine and starvation have struck only a few regions of the globe. But land, more than any other resource we deal with, is finite, and the expansion cannot continue indefinitely. Future gains in agricultural production are most probably going to come through more intensive use of existing cropland, heavier applications of fertilizers and other agricultural chemicals, and genetically-engineered crops requiring heavier applications of energy and water, than from an increase in the amount of the world's cropland.

– 78 –

Map 56 Annual Change in Forest Cover, 1981–1995

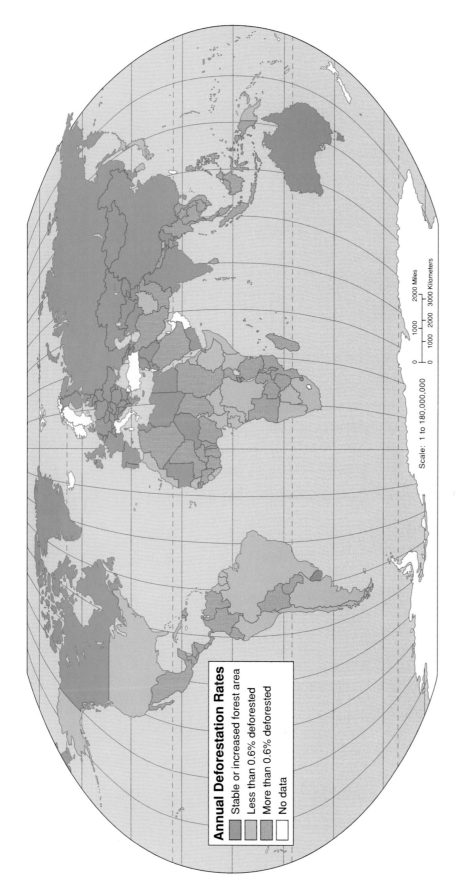

Annual Deforestation Rates

- Stable or increased forest area
- Less than 0.6% deforested
- More than 0.6% deforested
- No data

Scale: 1 to 180,000,000

0   1000   2000 Miles
0   1000   2000   3000 Kilometers

One of the most discussed environmental problems is that of deforestation. For most people, deforestation means clearing of tropical rain forests for agricultural purposes. Yet nearly as much forest land per year—much of it in North America, Europe, and Russia—is impacted by commercial lumbering as is cleared by tropical farmers and ranchers. Even in the tropics, much of the forest clearance is undertaken by large corporations producing high-value tropical hardwoods for the global market in furniture, ornaments, and other fine wood products. Still, it is the agriculturally driven clearing of the great rain forests of the Amazon Basin, west and central Africa, Middle America, and Southeast Asia that draws public attention. Although much concern over forest clearance focuses on the relationship between forest clearance and the reduction in the capacity of the world's vegetation system to absorb carbon dioxide (and thus delay global warming), of just as great concern are issues having to do with the loss of biodiversity (large numbers of plants and animals), the near-total destruction of soil systems, and disruptions in water supply that accompany clearing.

Map 57 The Loss of Biodiversity: Globally Threatened Animal Species

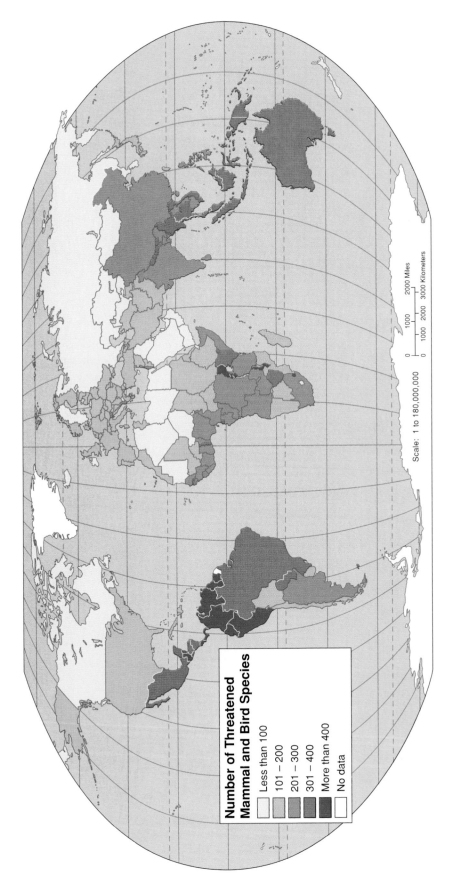

Scale: 1 to 180,000,000

Number of Threatened Mammal and Bird Species

- Less than 100
- 101 – 200
- 201 – 300
- 301 – 400
- More than 400
- No data

Threatened species are those in grave danger of going extinct. Their populations are becoming restricted in range, and the size of the populations required for sustained breeding is nearing a critical minimum. *Endangered species* are in immediate danger of becoming extinct. Their range is already so reduced that the animals may no longer be able to move freely within an ecozone, and their populations are at the level where the species may no longer be able to sustain breeding. Most species become threatened first and then endangered as their range and numbers continue to decrease. When people think of animal extinction, they think of large herbivorous species like the rhinoceros or fierce carnivores like lions, tigers, or grizzly bears. Certainly these animals make almost any list of endangered or threatened species. But there are literally hundreds of less conspicuous animals that are equally threatened. Extinction is normally nature's way of informing a species that it is inefficient. But conditions in the late twentieth century are controlled more by human activities than by natural evolutionary processes. Species that are endangered or threatened fall into that category because, somehow, they are competing with us or with our domesticated livestock for space and food. And in that competition the animals are always going to lose.

– 80 –

Map 58 The Loss of Biodiversity: Globally Threatened Plant Species

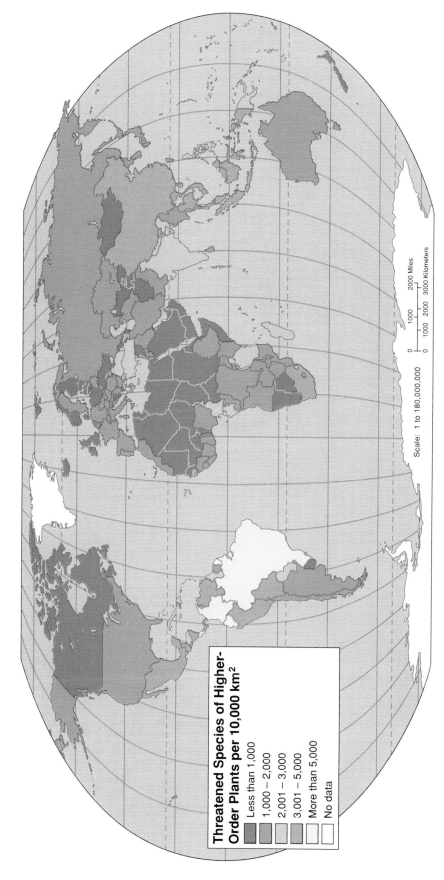

Threatened Species of Higher-Order Plants per 10,000 km²

- Less than 1,000
- 1,000 – 2,000
- 2,001 – 3,000
- 3,001 – 5,000
- More than 5,000
- No data

Scale: 1 to 180,000,000

0   1000   2000 Miles

0   1000   2000   3000 Kilometers

While most people tend to be more concerned about the animals on threatened and endangered species lists, the fact is that many more plants are in jeopardy, and the loss of plant life is, in all ecological regions, a more critical occurrence than the loss of animal populations. Plants are the primary producers in the ecosystem; that is, plants produce the food upon which all other species in the food web, including human beings, depend for sustenance. It is plants from which many of our critical medicines come, and it is plants that maintain the delicate balance between soil and water in most of the world's regions. When biogeographers and other environ-

mental scientists speak of a loss of "biodiversity," what they are most often describing is a loss of the richness and complexity of plant life that lends stability to ecosystems. Systems with more plant life tend to be more stable than those with less. For these and other reasons, the scientific concern over extinction is greater when applied to plants than to animals. It is difficult for people to become as emotional over a teak tree as they would over an elephant. But as great a tragedy as the loss of the elephant would be, the loss of the teak tree would be greater.

Map [59] The Risks of Desertification

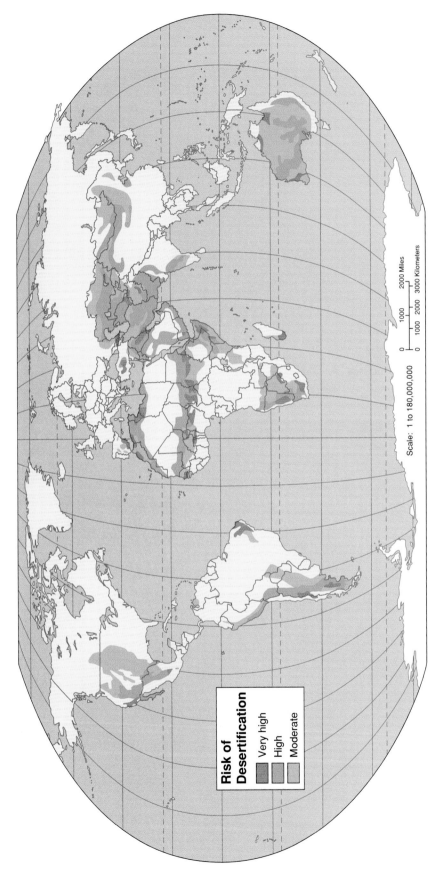

Scale: 1 to 180,000,000

0   1000   2000 Miles

0   1000   2000   3000 Kilometers

Risk of Desertification

Very high

High

Moderate

The awkward-sounding term "desertification" refers to a reduction in the food-producing capacity of drylands through vegetation, soil, and water changes that culminate in either a drier climate or in soil and plant systems that are less efficient in their use of water. Most of the world's existing drylands—the shortgrass steppes, the tropical savannas, the bunchgrass regions of the desert fringe—are fairly intensively used for agriculture and are, therefore, subject to the kinds of pressures that culminate in desertification. Most desertification is a natural process that occurs near the margins of desert regions. It is caused by dehydration of the soil's surface layers during periods of drought and by high water loss through evaporation in an environment of high temperature and high winds. This natural process

is greatly enhanced by human agricultural activities that expose topsoil to wind and water erosion. Among the most important practices that cause desertification are (1) overgrazing of rangelands, resulting from too many livestock on too small an area of land; (2) improper management of soil and water resources in irrigation agriculture, leading to accelerated erosion and to salt buildup in the soil; (3) cultivation of marginal terrain with soils and slopes that are unsuitable for farming; (4) surface disturbances of vegetation (clearing of thorn scrub, mesquite, chaparral, and similar vegetation) without soil protection efforts being made or replanting being done; and (5) soil compaction by agricultural implements, domesticated livestock, and rain falling on an exposed surface.

Map 60 Global Soil Degradation

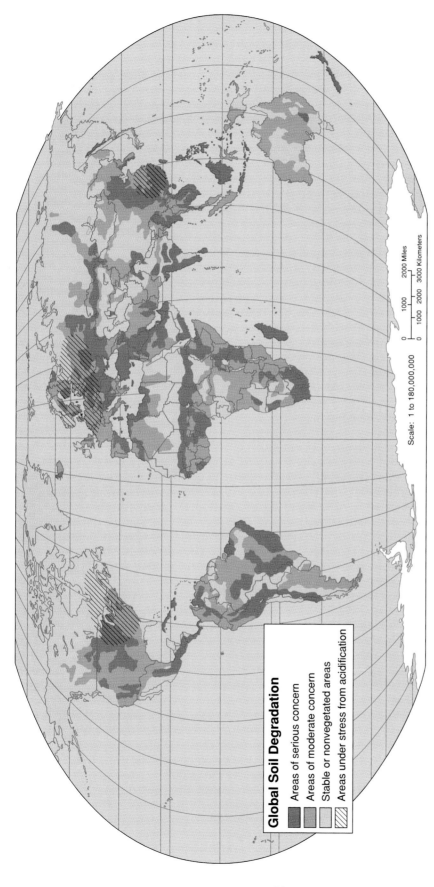

Global Soil Degradation

- Areas of serious concern
- Areas of moderate concern
- Stable or nonvegetated areas
- Areas under stress from acidification

Scale: 1 to 180,000,000

0   1000   2000 Miles

0   1000   2000   3000 Kilometers

Recent research has shown that more than 3 billion acres of the world's surface suffer from serious soil degradation, with more than 22 million acres so severely eroded or poisoned with chemicals that they can no longer support productive crop agriculture. Most of this soil damage has been caused by poor farming practices, overgrazing of domestic livestock, and deforestation. These activities strip away the protective cover of natural vegetation forests and grasslands, allowing wind and water erosion to remove the topsoil that contains necessary nutrients and soil microbes for plant growth. But millions of acres of topsoil have been degraded by chemicals as well. In some instances these chemicals are the result of overapplication of fertilizers, herbicides, pesticides, and other agricultural chemicals. In other instances, chemical deposition from industrial and urban wastes and from acid precipitation has poisoned millions of acres of soil. As the map shows, soil erosion and pollution are not problems just in developing countries with high population densities and increasing use of marginal lands. They also afflict the more highly developed regions of mechanized, industrial agriculture. While many methods for preventing or reducing soil degradation exist, they are seldom used because of ignorance, cost, or perceived economic inefficiency.

Map 61 Degree of Human Disturbance

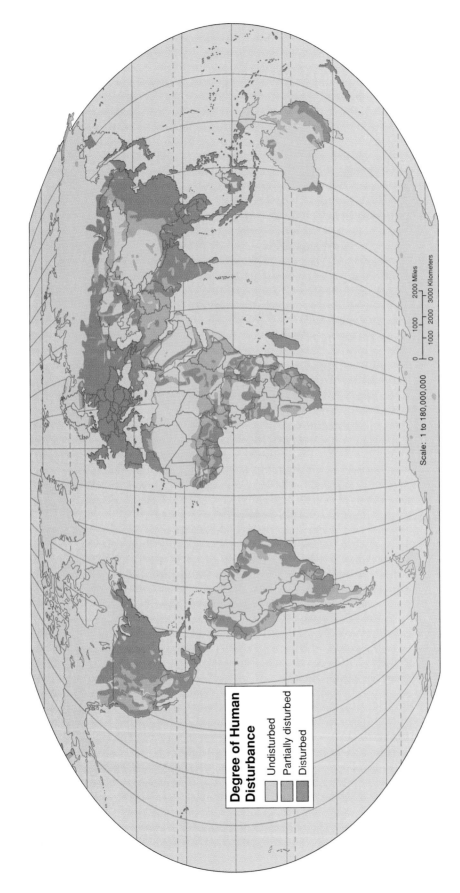

Scale: 1 to 180,000,000

0   1000   2000 Miles
0 1000 2000 3000 Kilometers

Degree of Human Disturbance

- Undisturbed
- Partially disturbed
- Disturbed

The data on human disturbance have been gathered from a wide variety of sources, some of them conflicting and not all of them reliable. Nevertheless, at a global scale this map fairly depicts the state of the world in terms of the degree to which humans have modified its surface. The undisturbed areas, covered with natural vegetation, generally have population densities under 10 persons per square mile. These areas are, for the most part, in the most inhospitable parts of the world: too high, too dry, too cold for permanent human habitation in large numbers. The partially disturbed areas are normally agricultural areas, either subsistence (such as shifting cultivation) or extensive (such as livestock grazing). They often contain ar-

eas of secondary vegetation, regrown after removal of original vegetation by humans. They are also often marked by a density of livestock in excess of carrying capacity, leading to overgrazing, which further alters the condition of the vegetation. The disturbed areas are those of permanent and intensive agriculture and urban settlement. The primary vegetation of these regions has been removed, with no evidence of regrowth or with current vegetation that is quite different from natural (potential) vegetation. Soils are in a state of depletion and degradation, and, in drier lands, desertification is a factor of human occupation. The disturbed areas match closely those areas of the world with the densest human populations.

Part VII

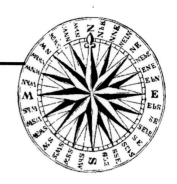

World Regions

Map 62 North America Political Map

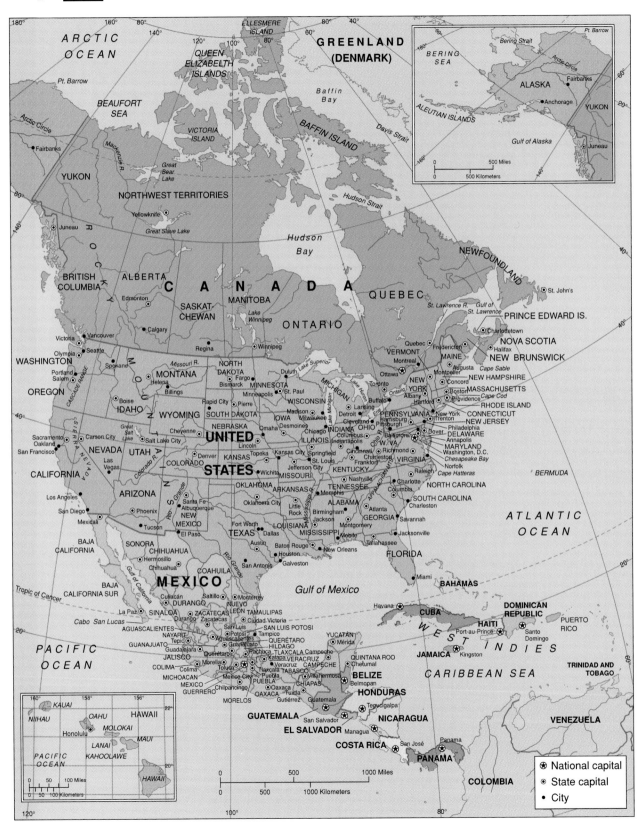

Map 63 North America Physical Map

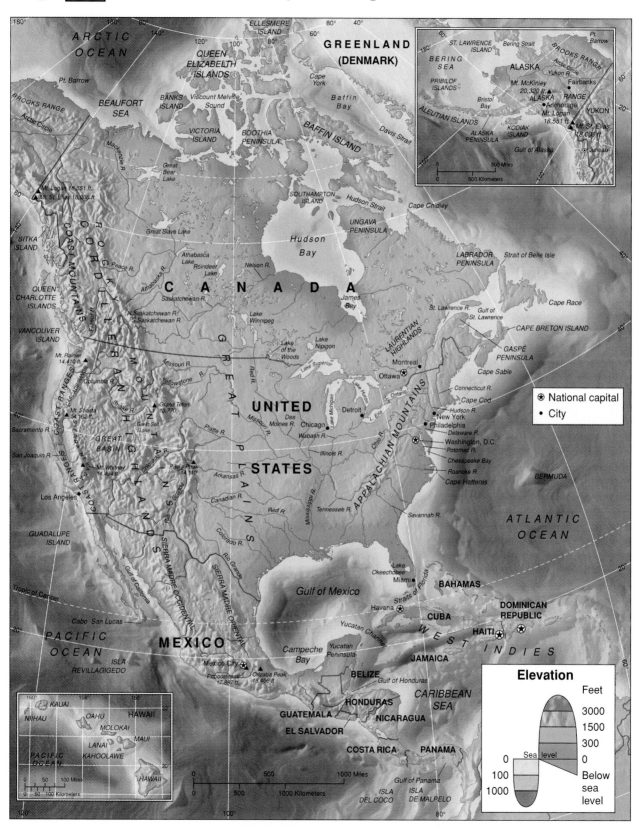

Elevation

Feet
3000
1500
300
Sea level 0
100 Below
1000 sea level

⊛ National capital
• City

North America Thematic Maps

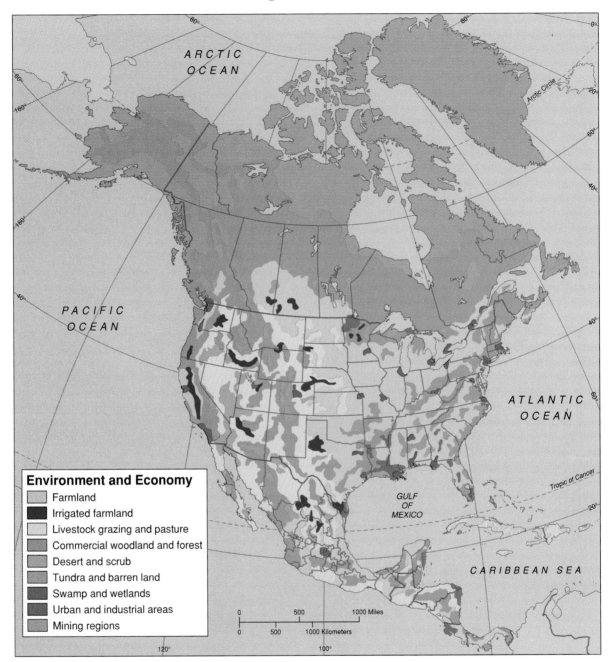

Environment and Economy

- Farmland
- Irrigated farmland
- Livestock grazing and pasture
- Commercial woodland and forest
- Desert and scrub
- Tundra and barren land
- Swamp and wetlands
- Urban and industrial areas
- Mining regions

Map ⟦64a⟧ Environment and Economy: The Use of Land

The use of land in North America represents a balance between agriculture, resource extraction, and manufacturing that is unmatched. The United States, as the world's leading industrial power, is also the world's leader in commercial agricultural production. Canada, despite its small population, is a ranking producer of both agricultural and industrial products and Mexico has begun to emerge from its developing nation status to become an important industrial and agricultural nation as well. The countries of Middle America and the Caribbean are just beginning the transition from agriculture to modern industrial economies. Part of the basis for the high levels of economic productivity in North America is environmental: a superb blend of soil, climate, and raw materials. But just as important is the cultural and social mix of the plural societies of North America, a mix that historically aided the growth of the economic diversity necessary for developed economies.

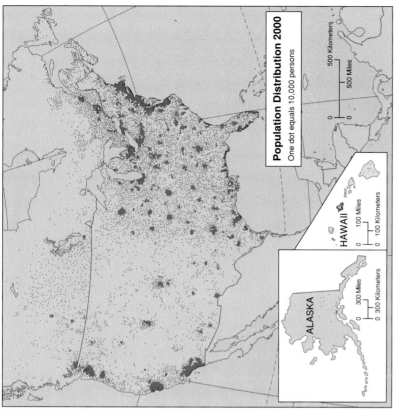

Map 64b Population Density

North America contains nearly 500 million people and the United States, with over 250 million inhabitants, is the third most populous country in the world, after China and India. Most of the present North American population has roots in the Old World. The native populations of the Americas had little or no resistance to Old World diseases in 1500 and within a couple of centuries of first contact with Europeans, most of the native peoples had either died out or preserved their genetic heritage by mixing with disease-resistant Old World populations. This left a North American population that is largely European, but with significant minorities resulting from the slave laborers imported from Africa, and from the mixture of native Americans with Europeans and/or Africans. The density of that population is largely the consequence of environmental factors (good soil, the availability of water, the presence of other resources) and cultural/economic ones (agricultural production, urbanization, industrialization).

Population Density
Persons per square mile (km)

- Uninhabited area
- Less than 2 (1)
- 2 – 25 (1–10)
- 25 – 50 (10 – 20)
- 50 – 150 (20 – 60)
- 150 – 300 (60 – 120)
- More than 300 (120)

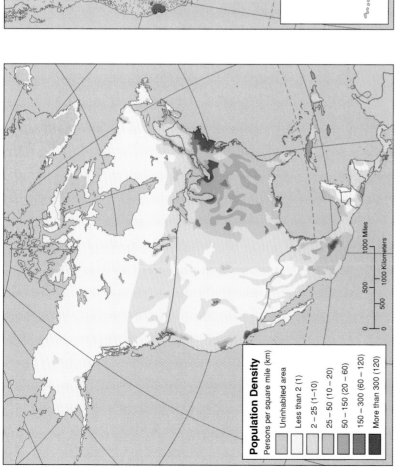

Population Distribution 2000
One dot equals 10,000 persons

ALASKA
0   300 Miles
0   300 Kilometers

HAWAII
0   100 Miles
0   100 Kilometers

0   500 Kilometers
0   500 Miles

Map 64c Population Distribution

Although population clustering is characteristic of highly economically developed regions, the Anglo-American population exhibits a remarkably clustered pattern. The primary reasons for this remarkable development are agricultural technology and affluence. A highly developed agricultural technology allows a small number of farmers, using sophisticated machinery, to grow and harvest enormous quantities of agricultural produce on large farms. In the United States, the world's leader in commercial agricultural production, only 2% of the population are farmers and that population is thinly distributed over wide areas. The vast majority of Americans live and work in city-suburb systems in which the widespread availability of private automobiles allows people to live considerable distances from where they work, keeping overall urban population densities relatively low but allowing for extensive urbanization—with cities large enough in area to be visible as population clusters on maps at this scale.

Map 65 South America Political Map

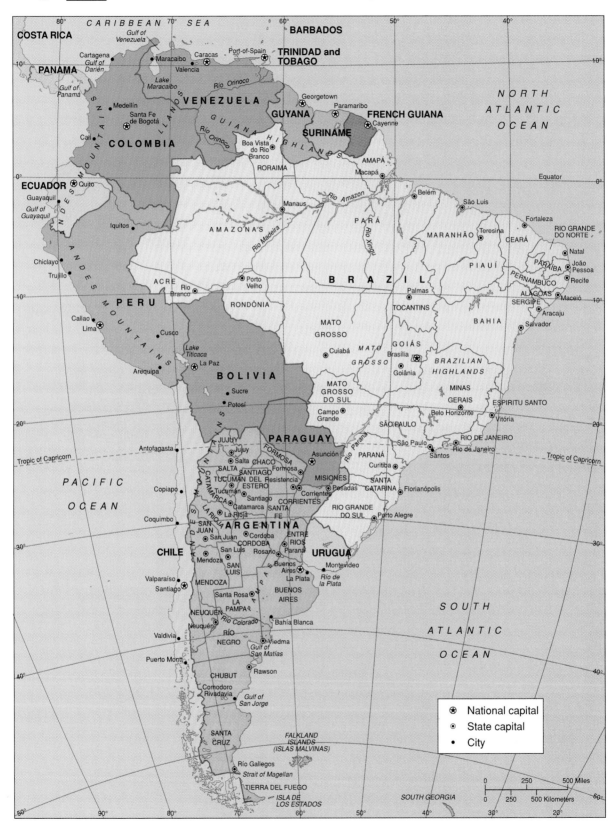

Map 65 — South America Political Map

CARIBBEAN SEA
80°   70°   60°   50°   40°

COSTA RICA
BARBADOS
Gulf of Venezuela
Cartagena
Gulf of Darién
• Maracaibo
Caracas •
Port-of-Spain
TRINIDAD and TOBAGO
PANAMA
Valencia •
Lake Maracaibo
Río Orinoco
Gulf of Panamá
NORTH ATLANTIC OCEAN
• Medellín
VENEZUELA
Georgetown
Santa Fe de Bogotá
GUIANA HIGHLANDS
GUYANA
Paramaribo
FRENCH GUIANA
Cali •
Río Orinoco
SURINAME
Cayenne
COLOMBIA
Boa Vista do Rio Branco
AMAPÁ
RORAIMA
Macapá
ECUADOR Quito
Equator
Guayaquil
Belém •
São Luís •
Gulf of Guayaquil
Rio Amazon
Manaus •
PARÁ
Fortaleza •
RIO GRANDE DO NORTE
• Iquitos
AMAZONAS
MARANHÃO
Teresina •
CEARÁ
Río Madeira
Rio Xingu
Natal
PARAÍBA
João Pessoa
• Chiclayo
BRAZIL
PIAUÍ
Recife
• Trujillo
ACRE
Rio Branco
Porto Velho •
PERNAMBUCO
PERU
RONDÔNIA
Palmas •
ALAGOAS
Maceió
SERGIPE
Aracaju
TOCANTINS
BAHIA
Callao
Lima
MATO GROSSO
Salvador •
• Cusco
MATO GROSSO
GOIÁS
Lake Titicaca
La Paz •
Cuiabá •
Brasília
BRAZILIAN HIGHLANDS
Arequipa •
Goiânia •
BOLIVIA
MINAS GERAIS
ESPIRITU SANTO
Sucre •
MATO GROSSO DO SUL
Belo Horizonte •
Vitória •
• Potosí
Campo Grande •
SÃO PAULO
RIO DE JANEIRO
Tropic of Capricorn
Antofagasta •
JUJUY
PARAGUAY
São Paulo •
Rio de Janeiro •
Tropic of Capricorn
• Jujuy
FORMOSA
Asunción
PARANÁ
Santos •
PACIFIC OCEAN
• Salta
SALTA
CHACO
Formosa
Curitiba •
SANTIAGO DEL ESTERO
Resistencia
MISIONES
SANTA CATARINA
Florianópolis •
TUCUMÁN
Corrientes •
Posadas •
Copiapo •
• Tucumán
CATAMARCA
Santiago •
CORRIENTES
SANTA FE
RIO GRANDE DO SUL
Porto Alegre •
LA RIOJA
Catamarca •
Coquimbo •
SAN JUAN
ARGENTINA
URUGUAY
La Rioja •
CÓRDOBA
• Córdoba
ENTRE RÍOS
CHILE
San Juan •
Montevideo
Valparaíso •
San Luis •
Rosario •
Paraná •
Mendoza •
SAN LUIS
Buenos Aires
Santiago •
MENDOZA
La Plata •
Río de la Plata
Santa Rosa •
LA PAMPA
BUENOS AIRES
SOUTH ATLANTIC OCEAN
NEUQUÉN
Bahía Blanca •
Neuquén •
Río Colorado
Valdivia •
RÍO NEGRO
Viedma •
Puerto Montt •
Gulf of San Matías
Rawson •
CHUBUT
Comodoro Rivadavia •
Gulf of San Jorge
SANTA CRUZ
FALKLAND ISLANDS (ISLAS MALVINAS)
Río Gallegos •
Strait of Magellan
TIERRA DEL FUEGO
ISLA DE LOS ESTADOS
SOUTH GEORGIA

	National capital
	State capital
	City

0   250   500 Miles
0   250   500 Kilometers

Map 66 South America Physical Map

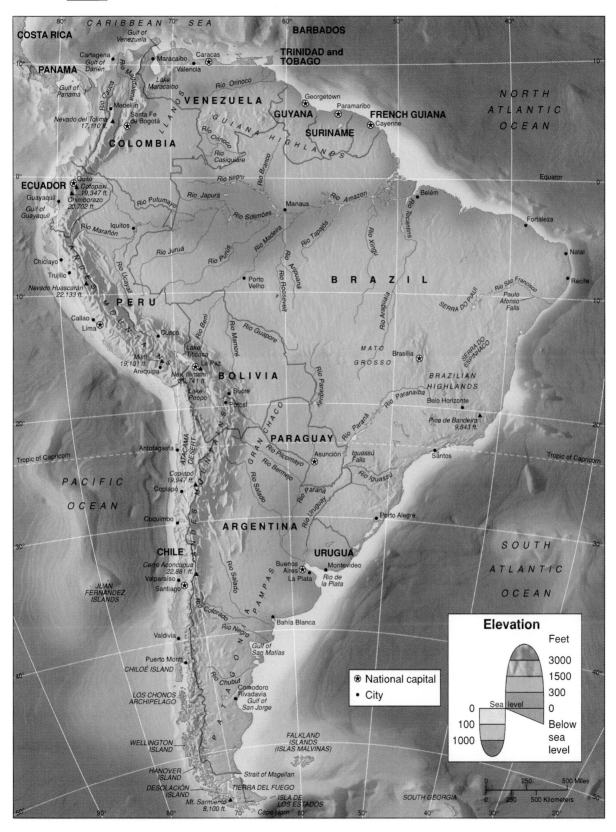

CARIBBEAN SEA

COSTA RICA

PANAMA

Gulf of Venezuela

Gulf of Darién

Gulf of Panamá

BARBADOS

TRINIDAD and TOBAGO

Cartagena

Maracaibo

Caracas

Valencia

Rio Magdalena

Rio Cauca

Medellín

Nevado del Tolima
17,110 ft.

Santa Fe de Bogotá

Lake Maracaibo

Rio Orinoco

VENEZUELA

LLANOS

GUIANA HIGHLANDS

Rio Orinoco

Rio Casiquiare

Georgetown

GUYANA

Paramaribo

SURINAME

FRENCH GUIANA

Cayenne

NORTH ATLANTIC OCEAN

COLOMBIA

Equator

ECUADOR

Quito

Cotopaxi
19,347 ft.

Chimborazo
20,702 ft.

Guayaquil

Gulf of Guayaquil

Rio Putumayo

Rio Japurá

Rio Negro

Rio Branco

Rio Solimões

Manaus

Rio Amazon

Belém

Fortaleza

Iquitos

Rio Marañón

Rio Juruá

Rio Purús

Rio Madeira

Rio Tapajós

Rio Tocantins

Natal

Chiclayo

Trujillo

Nevado Huascarán
22,133 ft.

Rio Ucayali

Porto Velho

Rio Roosevelt

Rio Xingu

B R A Z I L

Recife

Rio São Francisco

PERU

A N D E S M O U N T A I N S

Rio Beni

Rio Mamoré

Rio Guaporé

Rio Araguaia

SERRA DO PIAUÍ

Paulo Afonso Falls

Callao

Lima

Cusco

La Paz

Lake Titicaca

MATO GROSSO

Brasília

SERRA DO ESPINHAÇO

Misti
19,101 ft.

Arequipa

Nev. Illimani
20,741 ft.

Lake Poopó

BOLIVIA

Sucre

Potosí

Rio Paraguay

BRAZILIAN HIGHLANDS

Belo Horizonte

Pico de Bandeira
9,843 ft.

Antofagasta

ATACAMA DESERT

GRAN CHACO

PARAGUAY

Rio Pilcomayo

Asunción

Iguassú Falls

Santos

Tropic of Capricorn

PACIFIC OCEAN

Copiapó
19,947 ft.

Copiapó

Rio Bermejo

Rio Salado

Rio Paraná

Rio Paraná

Rio Iguassú

SOUTH ATLANTIC OCEAN

Coquimbo

ARGENTINA

Rio Paraná

Rio Uruguay

Porto Alegre

CHILE

Cerro Aconcagua
22,861 ft.

Valparaíso

Santiago

JUAN FERNÁNDEZ ISLANDS

Rio Salado

PAMPAS

URUGUAY

Buenos Aires

La Plata

Montevideo

Rio de la Plata

Rio Colorado

Bahía Blanca

Valdivia

Rio Negro

Gulf of San Matías

Puerto Montt

CHILOÉ ISLAND

Rio Chubut

Comodoro Rivadavia

Gulf of San Jorge

LOS CHONOS ARCHIPELAGO

WELLINGTON ISLAND

HANOVER ISLAND

FALKLAND ISLANDS
(ISLAS MALVINAS)

DESOLACIÓN ISLAND

Mt. Sarmiento
8,100 ft.

TIERRA DEL FUEGO

Strait of Magellan

ISLA DE LOS ESTADOS

Cape Horn

SOUTH GEORGIA

Elevation

⊗ National capital
• City

Feet
3000
1500
300
Sea level 0
Below sea level

100

1000

0   250   500 Miles

0   250   500 Kilometers

South America Thematic Maps

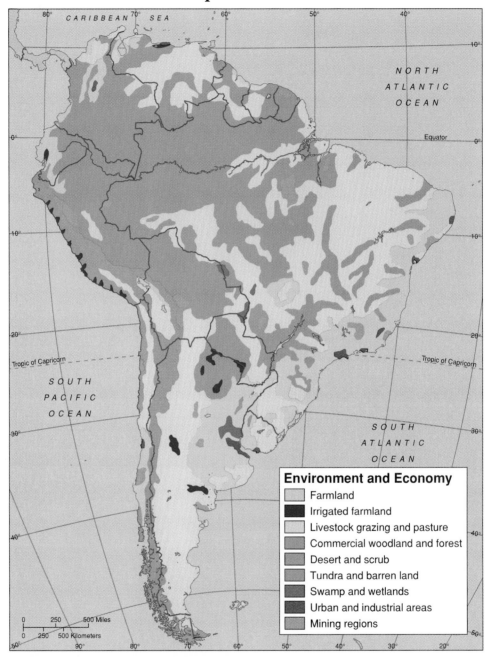

Environment and Economy

- Farmland
- Irrigated farmland
- Livestock grazing and pasture
- Commercial woodland and forest
- Desert and scrub
- Tundra and barren land
- Swamp and wetlands
- Urban and industrial areas
- Mining regions

Map 67a Environment and Economy

South America is a region just beginning to emerge from a colonial-dependency economy in which raw materials flowed from the continent to more highly-developed economic regions. With the exception of Brazil, Argentina, Chile, and Uruguay, most of the continent's countries still operate under the traditional mode of exporting raw materials in exchange for capital that tends to accumulate in the pockets of a small percentage of the population. The land use patterns of the continent are, therefore, still dominated by resource extraction and agriculture. A problem posed by these patterns is that little of the continent's land area is actually suitable for either commercial forestry or commercial crop agriculture without extremely high environmental costs. Much of the agriculture, then, is based on high value tropical crops that can be grown in small areas profitably, or on extensive livestock grazing. Even within the forested areas of the Amazon Basin where forest clearance is taking place at unprecedented rates, much of the land use that replaces forest is grazing.

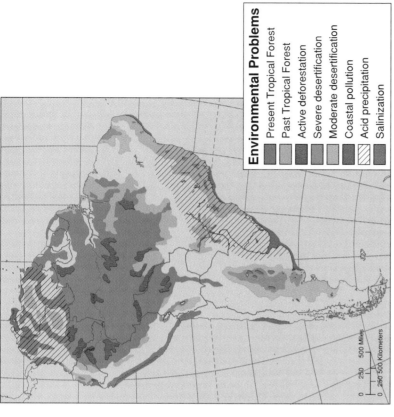

Map 67b Population Density

Population Density

Persons per square mile (kilometer)

- Uninhabited area
- Less than 2 (1)
- 2 – 50 (1 – 20)
- 50 – 150 (20 – 60)
- 150 – 300 (60 – 120)
- More than 300 (100)

Since so much of interior South America is uninhabitable (the high Andes) or only sparsely populated (the interior of the Amazon basin), the continent's population tends to be peripheral—approximately 90% of the continent's nearly 375 million people live within 150 miles of the sea. This population also tends to be heavily urbanized. Over 80% of South America's population lives in cities and the continent has three of the world's 15 largest cities—Rio de Janeiro, São Paulo, and Buenos Aires. São Paulo is the world's third largest urban agglomeration after Tokyo and New York. As in North America, most of the population of South America can trace at least part of its ancestry to the Old World. Throughout the Spanish-speaking parts of the continent the population is predominantly *mestizo* or mixed European and native South American. In Portuguese-speaking Brazil, in addition to a *mestizo* population, there is a significant admixture of African blood, the result of a large slave labor force imported from Africa to work the sugar, cotton, and other plantations of the colonial period.

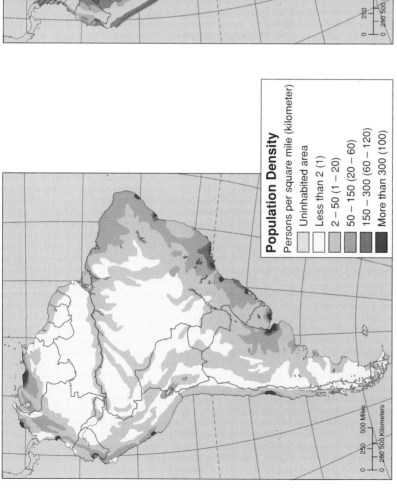

Environmental Problems

- Present Tropical Forest
- Past Tropical Forest
- Active deforestation
- Severe desertification
- Moderate desertification
- Coastal pollution
- Acid precipitation
- Salinization

Map 67c Environmental Problems

The drainage basin of the Amazon River and its tributaries, along with adjacent regions, is the world's largest remaining area of tropical forest. Much of the periphery of this vast forested region has already been cleared for farming and grazing and the ax and chainsaw and the flames are working their way steadily toward the interior. Tropical deforestation produces a loss of the biological diversity represented by the world's most biologically-productive ecosystem, along with changes in soil and soil-water systems. South America has other environmental problems: *desertification* in which grassland and/or scrub vegetation is converted to desert through overgrazing or other unwise agricultural practices; soil *salinization* in which soils become increasingly salty as the consequence of the over-application of irrigation water; *coastal and estuarine pollution* resulting from unregulated or unchecked use of coastal waters for industrial, commercial, and transportation purposes; and *acid precipitation* resulting from the combination of airborne industrial wastes and automobile-truck exhausts with water vapor to produce dry or wet acidic fallout.

Map 68 Europe Political Map

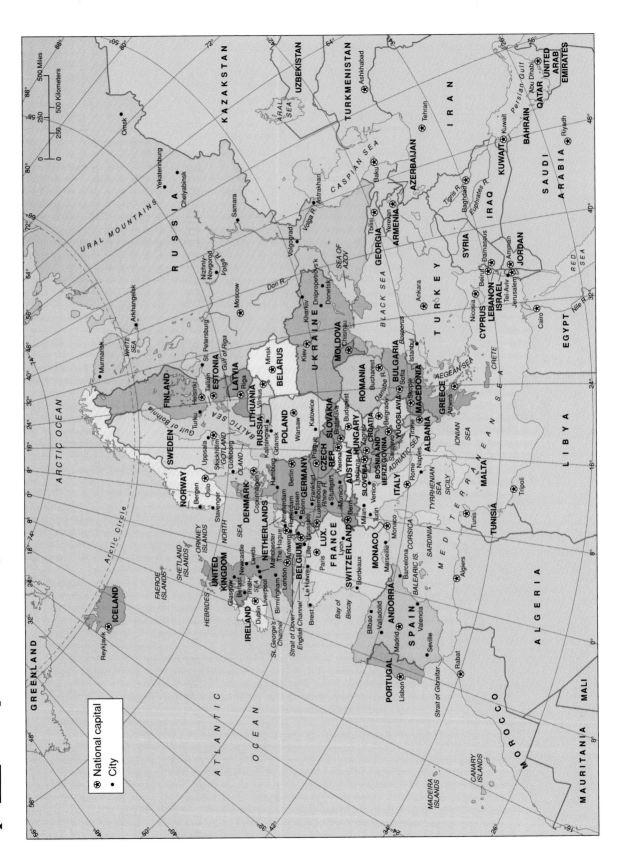

Map **69** Europe Physical Map

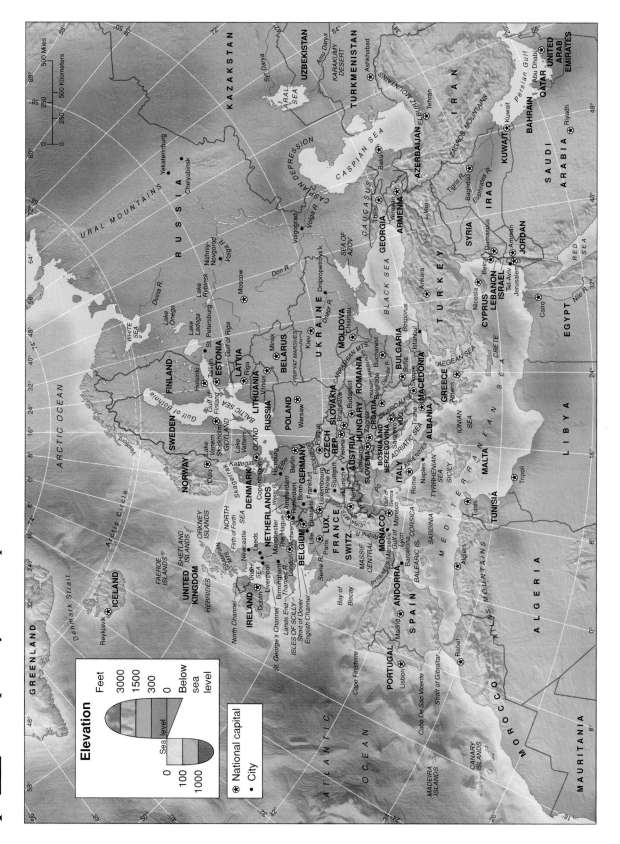

Elevation

Feet	
3000	
1500	
300	
0	Below sea level

Sea level

0
100
1000

⊛ National capital
• City

Europe Thematic Maps

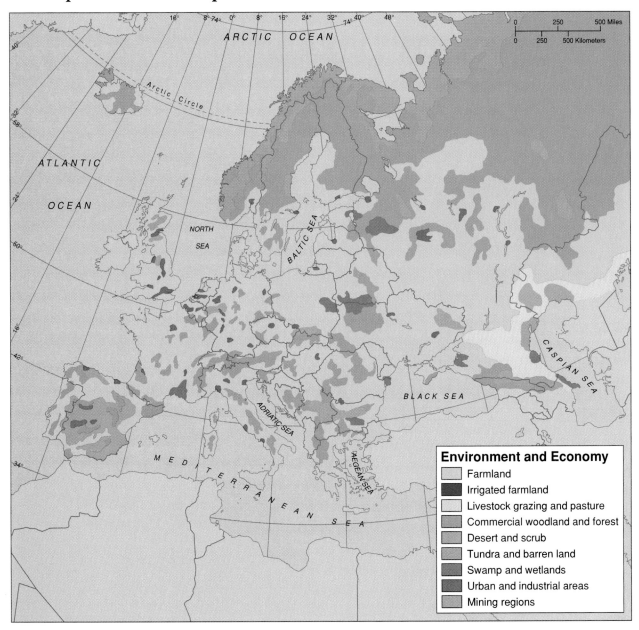

Environment and Economy

- Farmland
- Irrigated farmland
- Livestock grazing and pasture
- Commercial woodland and forest
- Desert and scrub
- Tundra and barren land
- Swamp and wetlands
- Urban and industrial areas
- Mining regions

Map 70a Environment and Economy: The Use of Land

More than any other continent, Europe bears the imprint of human activity—mining, forestry, agriculture, industry, and urbanization. Virtually all of western and central Europe's natural forest vegetation is gone, lost to clearing for agriculture beginning in prehistory, to lumbering that began in earnest during the Middle Ages, or more recently, to disease and destruction brought about by acid precipitation. Only in the far north and the east do some natural stands remain. The region is the world's most heavily industrialized and the industrial areas on the map represent only the largest and most significant. Not shown

are the industries that are found in virtually every small town and village and smaller city throughout the industrial countries of Europe. Europe also possesses abundant raw materials and a very productive agricultural base. The mineral resources have long been in a state of active exploitation and the mining regions shown on the map are, for the most part, old regions in upland areas that are somewhat less significant now than they may have been in the past. Agriculturally, the northern European plain is one of the world's great agricultural regions but most of Europe contains decent land for agriculture.

Map 70b Population Density

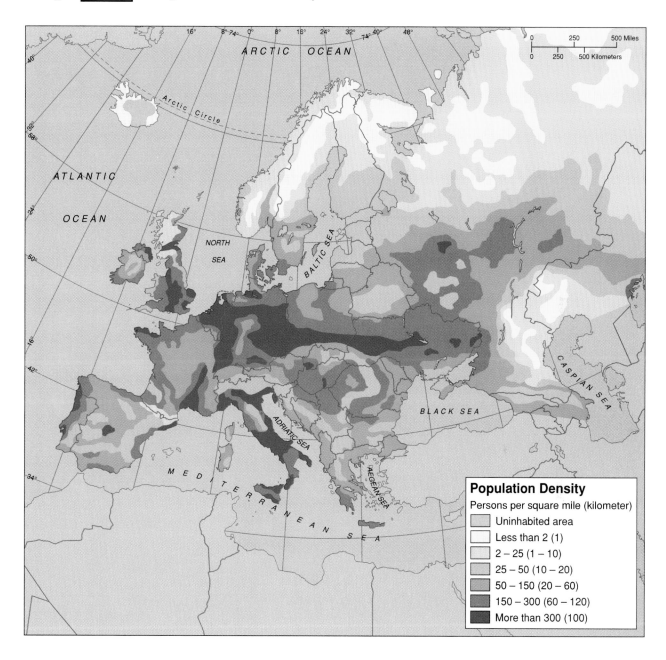

Population Density

Persons per square mile (kilometer)

- Uninhabited area
- Less than 2 (1)
- 2 – 25 (1 – 10)
- 25 – 50 (10 – 20)
- 50 – 150 (20 – 60)
- 150 – 300 (60 – 120)
- More than 300 (100)

Europe is one of the most densely settled regions of the world with an overall population density nearing 200 persons per square mile (80 per square kilometer), the consequence of a high level of urbanization and an economic system that is heavily industrialized. Even in agricultural regions, the population density is high. Beyond high density, the two chief identifying marks of the European population are remarkable diversity and unusual dynamics. For a small part of the world, Europe has cultural and ethnic diversity that is rarely matched elsewhere; more than 60 languages are spoken in an area not much larger than the United States. The population dynamics of Europe show a mature population that has passed through the "Demographic Transition"—a remarkable increase and then decline in growth rates resulting from the rise of an urban-industrial society. Only in other heavily industrialized regions of the world are found the very small overall growth rates characteristic of Europe.

Map 70c Political Changes in the 1990s

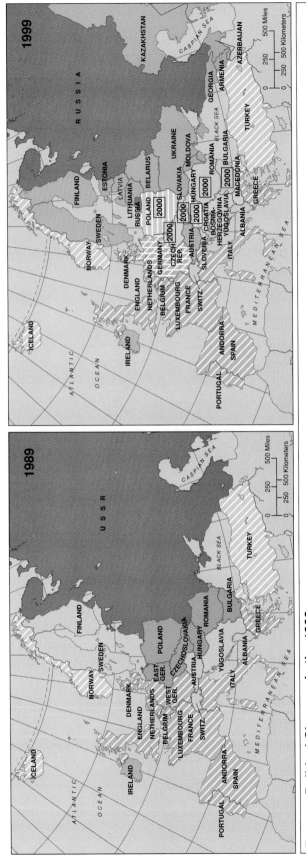

1999

Europe: Political Changes in the 1990s

1989

- Union of Soviet Socialist Republics
- Warsaw Pact Countries (excluding the USSR)
- North Atlantic Treaty Organization Countries
- European Community (formerly the EEC)

1999

- Former Republics of the USSR, now independent countries
- Russian Federation
- New European Countries resulting from reunification
- New European Countries resulting from partition
- NATO Countries, 1999
- Associated with NATO / Petitioned for entry
- European Union (formerly the EC)
- 2000 Year of projected entry into the European Union

During the last decade of the twentieth century, one of the most remarkable series of political geographic changes of the last 500 years took place. The bipolar "East-West" structure that had characterized Europe's political geography since the end of the Second World War altered in the space of a very few years. In the mid-1980s, as Soviet influence over eastern and central Europe weakened, those countries began to turn to the capitalist West. Between 1989, when the country of Hungary was the first Soviet satellite to open its borders to travel, and 1991, when the Soviet Union dissolved into 15 independent countries, abrupt change in political systems occurred. The result is a new map of Europe that includes a number of countries not present on the map of 1989. These countries have emerged as the result of reunification, separation, or independence from the old Soviet Union. The new political structure has been accompanied by growing economic cooperation.

Map 71 Africa Political Map

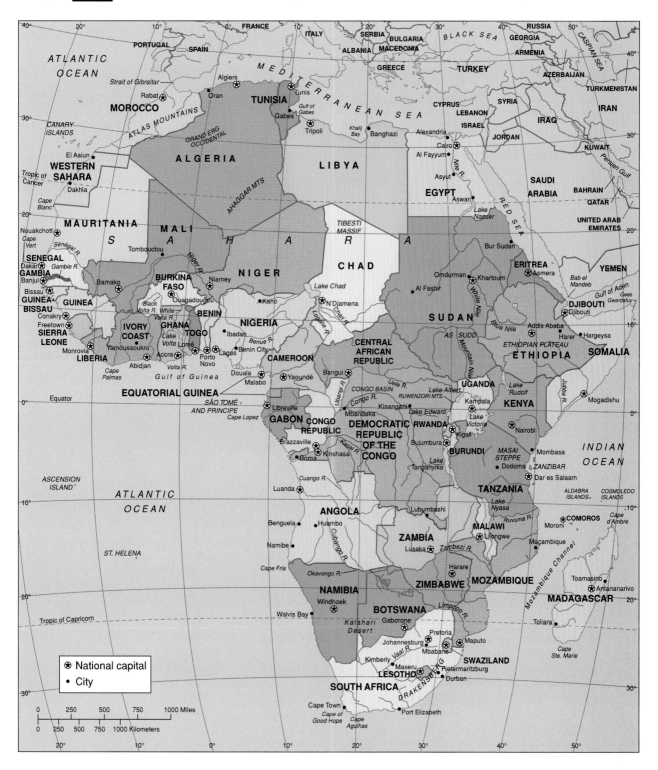

- ⊛ National capital
- • City

Map 72 Africa Physical Map

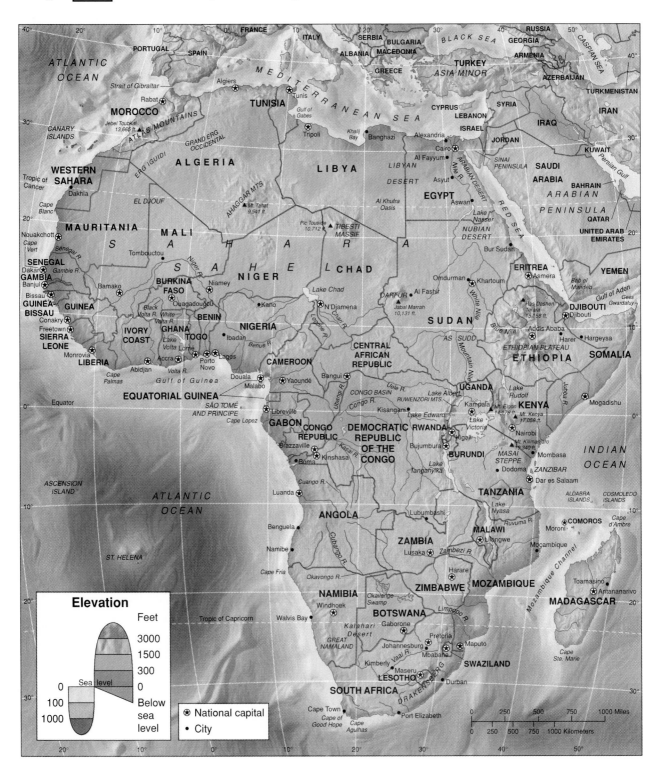

ATLANTIC OCEAN

Strait of Gibraltar

PORTUGAL
SPAIN
FRANCE
ITALY
SERBIA
ALBANIA
MACEDONIA
BULGARIA
GREECE
TURKEY
ASIA MINOR
RUSSIA
GEORGIA
ARMENIA
AZERBAIJAN
TURKMENISTAN
CASPIAN SEA
BLACK SEA

MEDITERRANEAN SEA

Algiers
Tunis
Rabat
MOROCCO
TUNISIA
Gulf of Gabes
Khalij Bay
Tripoli
Banghazi
Alexandria
Cairo
Al Fayyum
CYPRUS
LEBANON
SYRIA
ISRAEL
JORDAN
IRAQ
IRAN
KUWAIT
Persian Gulf

Jebel Toubkal 13,665 ft.
ATLAS MOUNTAINS
GRAND ERG OCCIDENTAL

CANARY ISLANDS
Tropic of Cancer
WESTERN SAHARA
Dakhla
Cape Blanc

EL DJOUF
ERG IGUIDI

ALGERIA
LIBYA
Al Khufra Oasis
LIBYAN DESERT
EGYPT
Asyut
Aswan
Lake Nasser
NUBIAN DESERT
Bur Sudan
RED SEA
SINAI PENINSULA
SAUDI ARABIA
BAHRAIN
QATAR
UNITED ARAB EMIRATES
ARABIAN PENINSULA
ARABIAN DESERT
Nile R.

MAURITANIA
MALI
AHAGGAR MTS.
Mt. Tahat 9,541 ft.
Pic Toussidé 10,712 ft.
TIBESTI MASSIF
SAHARA
SAHEL
NIGER
CHAD

Nouakchott
Cape Vert
SENEGAL
Dakar
GAMBIA
Banjul
Gambie R.
Senegal R.
Tombouctou
Bamako
BURKINA FASO
Niamey
Kano
N'Djamena
Lake Chad
DARFUR
Al Fashir
Jabal Marrah 10,131 ft.
SUDAN
Omdurman
Khartoum
ERITREA
Asmera
YEMEN
Bab el Mandeb
Gulf of Aden
Gees Gwardafuy
DJIBOUTI
Djibouti

BISSAU
GUINEA BISSAU
Conakry
Freetown
SIERRA LEONE
Monrovia
LIBERIA
GUINEA
Black Volta R.
White Volta R.
Ouagadougou
BENIN
GHANA
TOGO
Ibadan
Lome
Accra
Porto Novo
Lagos
NIGERIA
Benue R.
Logone R.
Chari R.
White Nile
Blue Nile
AS SUDD
Addis Ababa
Harer
Hargeysa
ETHIOPIAN PLATEAU
ETHIOPIA
SOMALIA
Ras Dashen Terara 15,158 ft.

IVORY COAST
Lake Volta
Volta R.
Abidjan
Cape Palmas
Gulf of Guinea

Equator

Cape Lopez
SÃO TOMÉ AND PRINCIPE
Libreville
GABON
EQUATORIAL GUINEA
Douala
Malabo
Yaoundé
CAMEROON
Bangui
CENTRAL AFRICAN REPUBLIC
Ubangi R.
Uele R.
CONGO BASIN
Congo R.
Kisangani
Lake Albert
RUWENZORI MTS.
Lake Edward
UGANDA
Kampala
Mt. Elgon 14,178 ft.
KENYA
Mt. Kenya 17,058 ft.
Lake Rudolf
Mogadishu

CONGO REPUBLIC
Brazzaville
Kinshasa
Boma
DEMOCRATIC REPUBLIC OF THE CONGO
RWANDA
Kigali
Bujumbura
BURUNDI
Lake Victoria
Lake Tanganyika
Nairobi
Mt. Kilimanjaro 19,340 ft.
MASAI STEPPE
Mombasa
Dodoma
ZANZIBAR
Dar es Salaam
INDIAN OCEAN
Juba R.
Kasai R.
Mountain Nile

ASCENSION ISLAND

ATLANTIC OCEAN

Luanda
Cuango R.
ANGOLA
Benguela
Namibe
Cubango R.
Lubumbashi
Lake Nyasa
Ruvuma R.
TANZANIA
ALDABRA ISLANDS
COSMOLEDO ISLANDS
Cape d'Ambre
COMOROS
Moroni

ST. HELENA

Cape Fria
Okavango R.
Okavango Swamp
NAMIBIA
Windhoek
ZAMBIA
Lusaka
Zambezi R.
Harare
ZIMBABWE
MALAWI
Lilongwe
Moçambique
MOZAMBIQUE
Mozambique Channel
Toamasino
Antananarivo
MADAGASCAR

Tropic of Capricorn
Walvis Bay
Kalahari Desert
GREAT NAMALAND
BOTSWANA
Gaborone
Limpopo R.
Okavango Swamp

Cape Fria

SWAZILAND
Mbabane
Maputo
Pretoria
Johannesburg
Kimberley
Vaal R.
Maseru
LESOTHO
Durban
SOUTH AFRICA
DRAKENSBERG
Cape Ste. Marie

Cape Town
Cape of Good Hope
Cape Agulhas
Port Elizabeth

Elevation

	Feet
	3000
	1500
	300
Sea level 0	0
100	Below sea level
1000	

⊛ National capital
• City

0   250   500   750   1000 Miles
0 250 500 750 1000 Kilometers

Africa Thematic Maps

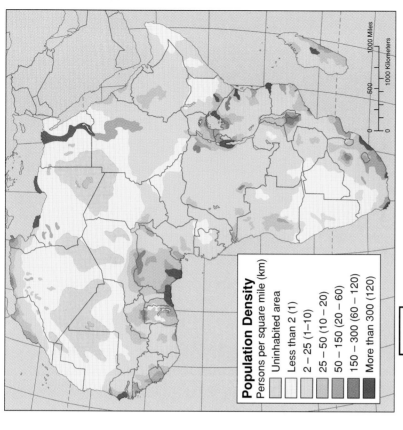

Map `73b` **Population Density**

Nearly 800 million people occupy the African continent, approximately 1/8th of the world's population. In general, this population has two chief characteristics: a low-level of quality of life and growth rates that are among the world's highest. On a continent beset by poverty, recurrent internal civil and tribal war, and a host of environmental problems, the populations of many African countries are nevertheless increasing at a rate above 3% per year. The bulk of the population is concentrated in relatively small areas of the Mediterranean coastal regions of the north, the bulge of West Africa, and the eastern coastal and highland zone stretching from South Africa to Kenya. Where populations in other parts of the world tend to avoid highland locations, in Africa highlands often tend to be the most densely settled regions: moister with better soils, freer of insect pests, and somewhat cooler.

Population Density
Persons per square mile (km)

- Uninhabited area
- Less than 2 (1)
- 2 – 25 (1–10)
- 25 – 50 (10 – 20)
- 50 – 150 (20 – 60)
- 150 – 300 (60 – 120)
- More than 300 (120)

1000 Miles
1000 Kilometers
0   500   1000

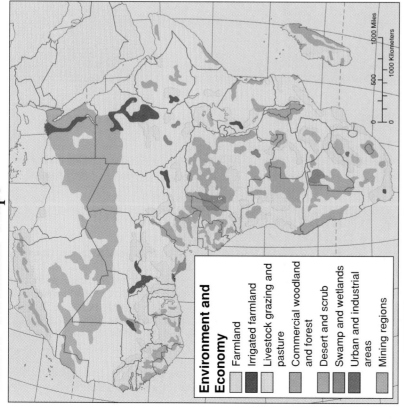

Map `73a` **Environment and Economy**

Africa's economic landscape is dominated by subsistence, or marginally-commercial agricultural activities and raw material extraction, engaging three-fourths of Africa's workers. Much of this grazing land is very poor desert scrub and bunch grass that is easily impacted by cattle, sheep, and goats. Growing human and livestock populations place enormous stress on this fragile support capacity and the result is desertification: the conversion of even the most minimal of grazing environments to desert. Balanced against the enormous reaches of land suitable only for grazing is a small quantity of land suitable for crop farming. Although the continent has approximately 20 percent of the world's total land area, the proportion of Africa's arable land is small. The agricultural environment is also uncertain; unpredictable precipitation and poor soils hamper crop agriculture.

Environment and Economy

- Farmland
- Irrigated farmland
- Livestock grazing and pasture
- Commercial woodland and forest
- Desert and scrub
- Swamp and wetlands
- Urban and industrial areas
- Mining regions

1000 Miles
1000 Kilometers
0   500   1000

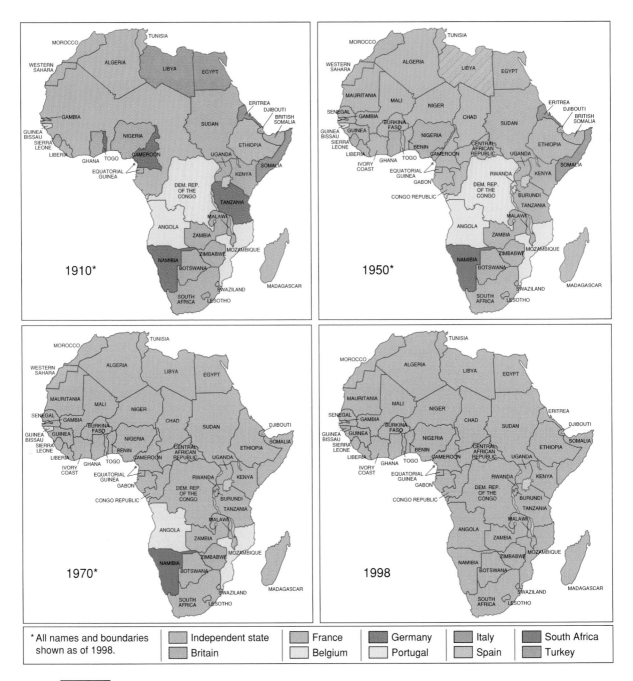

Map 73c **Colonialism to Independence: 1910 to 1998**

In few parts of the world has the transition from colonialism to independence been as abrupt as on the African continent. Most African states did not become colonies until the nineteenth century and did not become independent until the twentieth, nearly all of them after World War II. Much of the colonial legacy of the European colonial powers in Africa is social and economic. The African colony provided the mother country with raw materials in exchange for marginal economic returns and many African countries still exist in this colonial dependency relationship. An even more important component of the colonial legacy of Europe in Africa is geopolitical. When

the world's colonial powers joined at the Conference of Berlin in 1884, they divided up Africa to fit their own needs, drawing boundary lines on maps without regard for terrain or drainage features, or for tribal/ethnic linguistic, cultural, economic, or political borders. Traditional Africa was enormously disrupted by this process. After independence, African countries retained boundaries that are legacies of the colonial past and African countries today are beset by internal problems related to tribal and ethnic conflicts, the disruption of traditional migration patterns, and inefficient spatial structures of market and supply.

Map 74 Asia Political Map

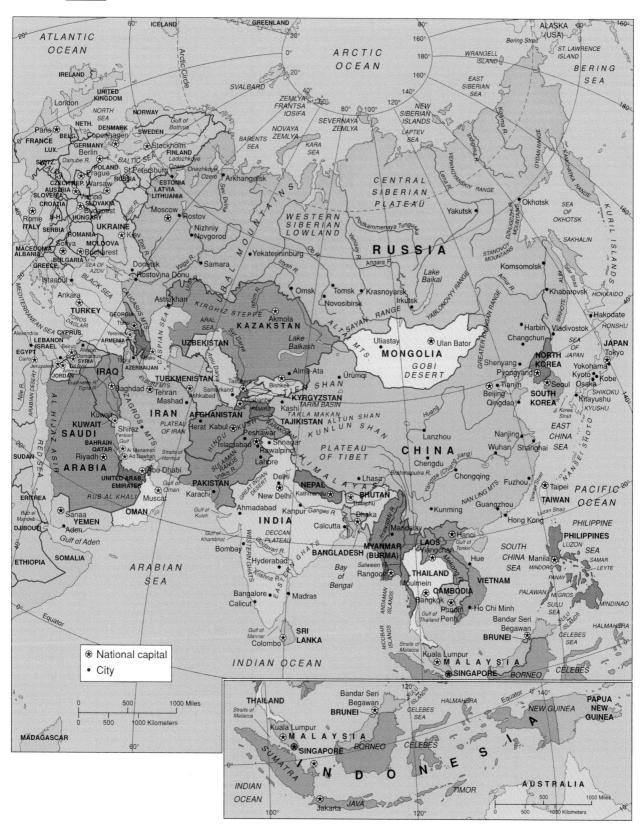

Map 75 Asia Physical Map

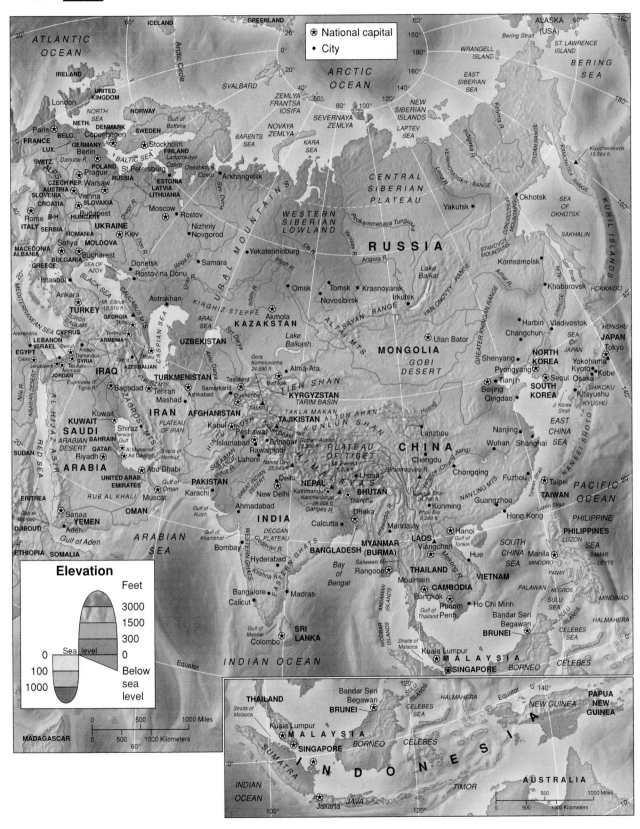

National capital ⊛
City •

Elevation

Feet
3000
1500
300
0 Sea level 0
100 Below
1000 sea level

Asia Thematic Maps

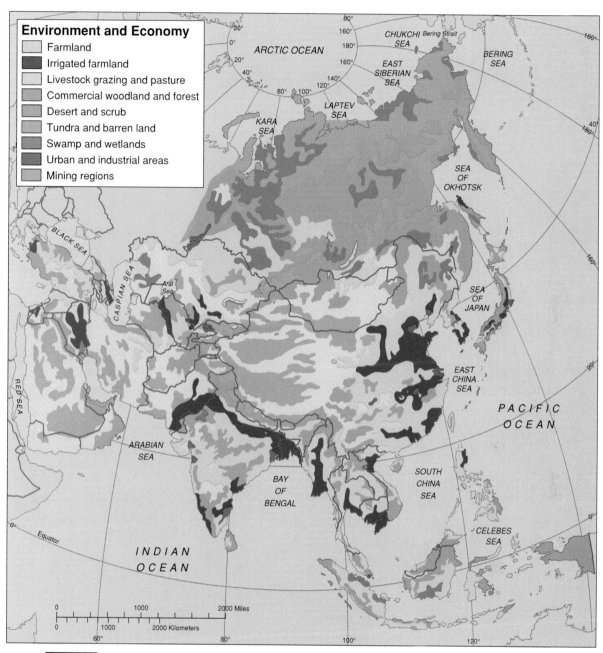

Environment and Economy

	Farmland
	Irrigated farmland
	Livestock grazing and pasture
	Commercial woodland and forest
	Desert and scrub
	Tundra and barren land
	Swamp and wetlands
	Urban and industrial areas
	Mining regions

Map 76a Environment and Economy

Asia is a land of extremes of land use with some of the world's most heavily industrialized regions, barren and empty areas, and productive and densely populated farm regions. Asia is a region of rapid industrial growth. Yet Asia remains an agricultural region with three out of every four workers engaged in agriculture. Asian commercial agriculture and intensive subsistence agriculture is characterized by irrigation. Some of Asia's irrigated lands are desert requiring additional water. But most of the Asian irrigated regions have sufficient precipitation for crop ag-

riculture and irrigation is a way of coping with seasonal drought—the wet-and-dry cycle of the monsoon—often gaining more than one crop per year on irrigated farms. Agricultural yields per unit area in many areas of Asia are among the world's highest. Because the Asian population is so large and the demands for agricultural land so great, Asia is undergoing rapid deforestation and some areas of the continent have only small remnants of a once-abundant forest reserve.

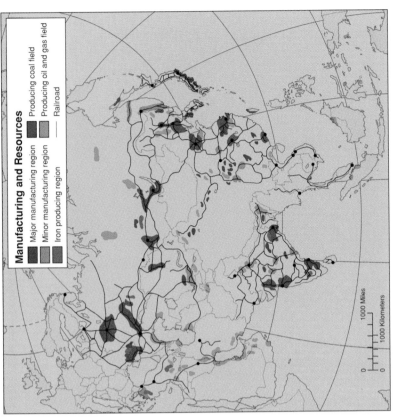

Map 76b Population Density

With one-third of the world's land area and nearly two-thirds of the world's population, Asia is more densely settled than any other region of the world. In some of the continent's farming regions, agricultural population density exceeds 2000 persons per square mile. In some portions of the continent, particularly the Islamic areas of Central and Southwest Asia, this already large population is growing very rapidly with some countries having population growth rates above 3% per year and doubling times between 20 and 25 years. The populations of neither China nor India are growing particularly rapidly but since both countries have population bases that are enormous, the absolute number of Indians and Chinese added to the world's population each year is staggering. In spite of these massive populations, Asia also contains areas that are either completely uninhabited or have population densities that are as low as any on earth.

Population Density

Persons per square mile (km)

- Uninhabited area
- Less than 2 (1)
- 2 – 25 (1–10)
- 25 – 50 (10 – 20)
- 50 – 150 (20 – 60)
- 150 – 300 (60 – 120)
- More than 300 (120)

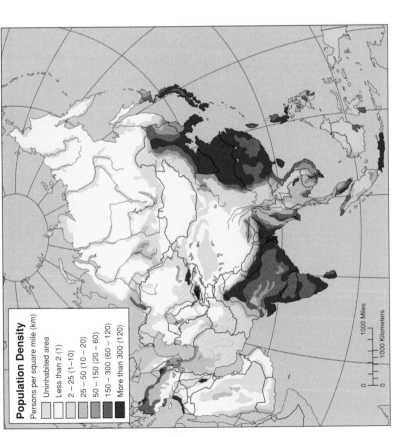

Manufacturing and Resources

- Major manufacturing region
- Minor manufacturing region
- Iron producing region
- Producing coal field
- Producing oil and gas field
- Railroad

Map 76c Industrialization: Manufacturing and Resources

International economists have predicted that the next century will be one of Asian economic dominance, as other Asian nations approach the economic levels of industrial giant Japan. Asian countries have begun to industrialize rapidly and have increased industrial production more than 100 percent in the last decade. But nothing guarantees that this success will continue. Before Asia can rival Europe and North America in industrial output, it must master the great distances separating critical raw material and production sites from the locations of the markets for them. Such a mastery is achieved only by the development of efficient transportation systems. Usually that means water travel and here Asia is remarkably deficient when compared with Europe and North America. Nevertheless, the mixed prospects for Asian economic growth into the next century are a great deal better than would have been predicted a decade ago.

Map 77 Australasia Political Map

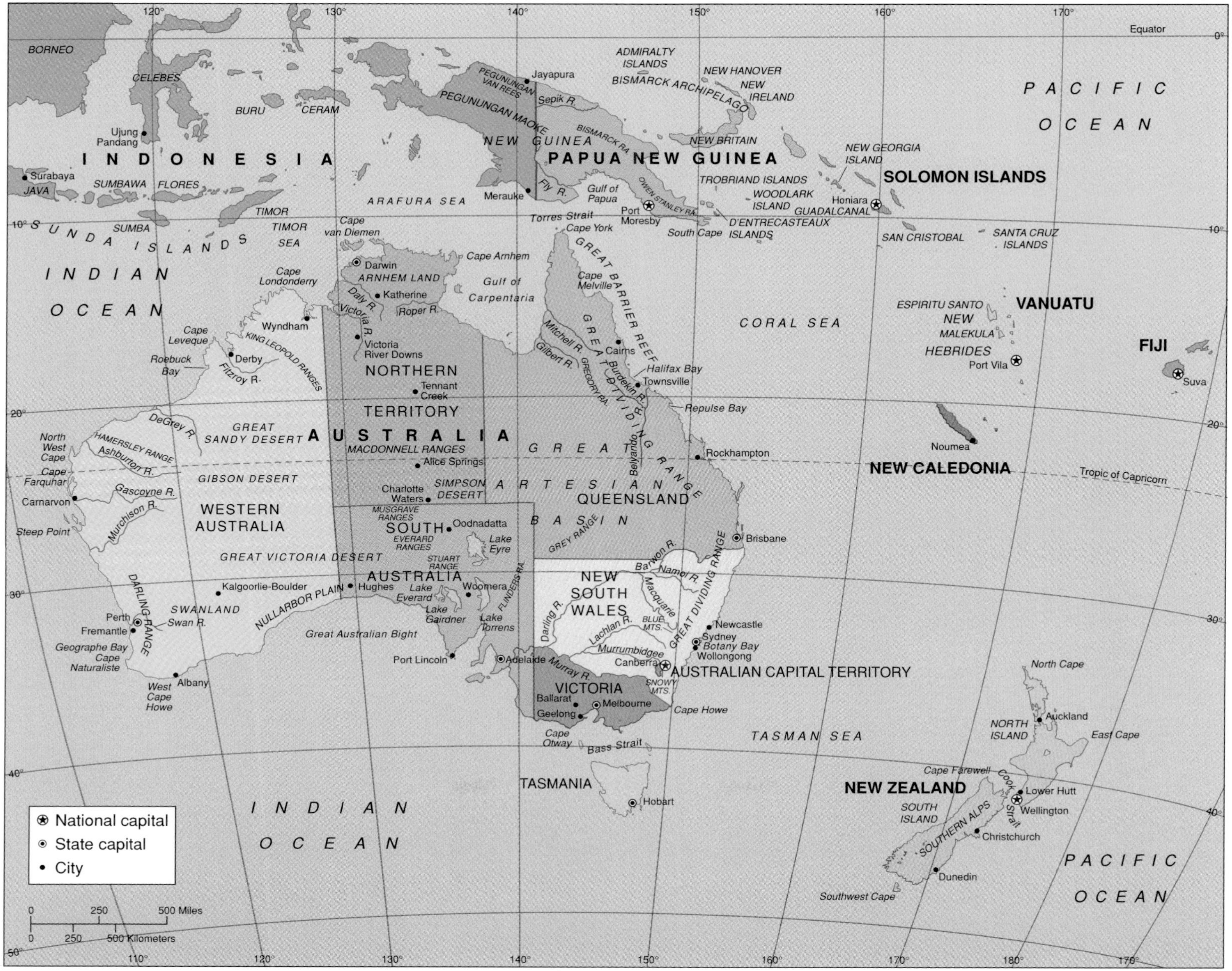

PACIFIC OCEAN

Equator

FIJI
Suva

SOLOMON ISLANDS
VANUATU
SANTA CRUZ ISLANDS
SAN CRISTOBAL
NEW GEORGIA ISLAND
Honiara
GUADALCANAL
WOODLARK ISLAND
D'ENTRECASTEAUX ISLANDS
TROBRIAND ISLANDS

ESPIRITU SANTO
NEW HEBRIDES
MALEKULA
Port Vila

NEW CALEDONIA
Nouméa

Tropic of Capricorn

NEW ZEALAND
NORTH ISLAND
SOUTH ISLAND
SOUTHERN ALPS
North Cape
Auckland
East Cape
Lower Hutt
Wellington
Cook Strait
Christchurch
Dunedin
Cape Farewell
Southwest Cape

PACIFIC OCEAN

TASMAN SEA

PAPUA NEW GUINEA
ADMIRALTY ISLANDS
BISMARCK ARCHIPELAGO
NEW HANOVER
NEW IRELAND
NEW BRITAIN
BISMARCK RA.
OWEN STANLEY RA.
Port Moresby
Gulf of Papua
South Cape
NEW GUINEA
Sepik R.
Fly R.
Merauke
Jayapura
PEGUNUNGAN MAOKE
PEGUNUNGAN VAN REES

INDONESIA
BORNEO
CELEBES
CERAM
BURU
Ujung Pandang
Surabaya
JAVA
SUMBAWA
SUMBA
FLORES
TIMOR
SUNDA ISLANDS
TIMOR SEA

ARAFURA SEA
Cape van Diemen
Cape Arnhem
Gulf of Carpentaria
Cape York
Cape Melville

CORAL SEA
GREAT BARRIER REEF

Torres Strait

NORTHERN TERRITORY
ARNHEM LAND
Darwin
Katherine
Daly R.
Roper R.
Victoria River Downs
Victoria R.
MACDONNELL RANGES
Tennant Creek
Alice Springs
SIMPSON DESERT

QUEENSLAND
GREAT DIVIDING RANGE
Cairns
Townsville
Halifax Bay
Repulse Bay
Burdekin R.
Mitchell R.
Gilbert R.
GREGORY RA.
Belyando R.
Rockhampton
GREAT ARTESIAN BASIN
Brisbane
GREY RANGE
Warrego R.
Barwon R.

WESTERN AUSTRALIA
GREAT SANDY DESERT
GIBSON DESERT
GREAT VICTORIA DESERT
KING LEOPOLD RANGES
HAMERSLEY RANGE
Wyndham
Derby
Fitzroy R.
Cape Leveque
Roebuck Bay
Cape Londonderry
De Grey R.
Ashburton R.
Gascoyne R.
Murchison R.
Carnarvon
North West Cape
Cape Farquhar
Steep Point
Geographe Bay
Cape Naturaliste
DARLING RANGE
Perth
Fremantle
Swan R.
Kalgoorlie-Boulder
NULLARBOR PLAIN
SWANLAND
West Cape Howe
Albany
Great Australian Bight

SOUTH AUSTRALIA
MUSGRAVE RANGES
EVERARD RANGES
STUART RANGE
FLINDERS RA.
Oodnadatta
Charlotte Waters
Lake Eyre
Lake Torrens
Lake Gairdner
Lake Everard
Woomera
Hughes
Port Lincoln
Port Augusta
Adelaide

NEW SOUTH WALES
Newcastle
Sydney
Botany Bay
Wollongong
Cape Howe
Macquarie R.
Murrumbidgee R.
Murray R.
Darling R.
Lachlan R.
BLUE MTS.
SNOWY MTS.

AUSTRALIAN CAPITAL TERRITORY
Canberra

VICTORIA
Melbourne
Ballarat
Geelong
Cape Otway

BASS STRAIT
TASMANIA
Hobart

INDIAN OCEAN

A U S T R A L I A

National capital
State capital
City

500 Miles
500 Kilometers
250
0

— 107 —

Map 78 Australasia Physical Map

Elevation

Feet	
3000	
1500	
300	
0	Sea level
Below sea level	

| 0 |
| 100 |
| 1000 |

⊛ National capital
◉ State capital
• City

500 Miles
500 Kilometers

Australasia Thematic Maps

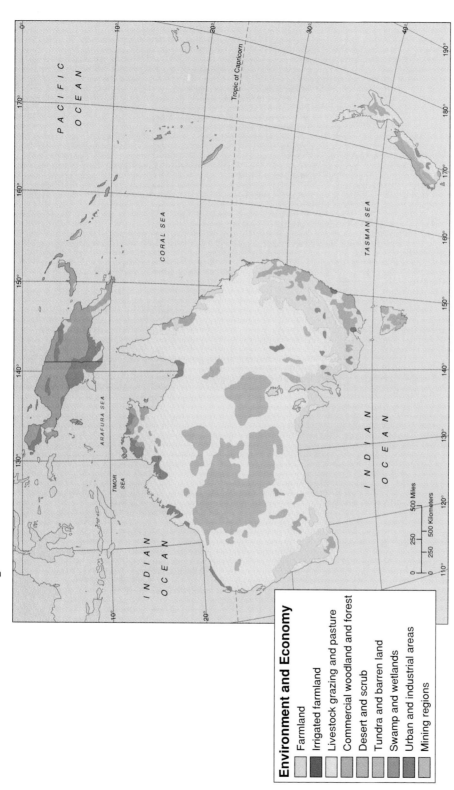

Environment and Economy

- Farmland
- Irrigated farmland
- Livestock grazing and pasture
- Commercial woodland and forest
- Desert and scrub
- Tundra and barren land
- Swamp and wetlands
- Urban and industrial areas
- Mining regions

Map 79a Environment and Economy

Australasia is dominated by the world's smallest and most uniform continent. Flat, dry, and mostly hot, Australia has the simplest of land use patterns: where rainfall exists so does agricultural activity. Two agricultural patterns dominate the map: livestock grazing, primarily sheep, and wheat farming, although some sugar cane production exists in the north and some cotton is grown elsewhere. Only about 6 percent of the continent consists of arable land so the areas of wheat farming, dominant as they may be in the context of Australian agriculture, are small. Australia also supports a healthy mineral resource economy, with iron and copper and precious metals making up the bulk of the extraction. Elsewhere in the region, tropical forests dominate Papua New Guinea, with some subsistence agriculture and livestock. New Zealand's temperate climate with abundant precipitation supports a productive livestock industry and little else besides tourism—which is an important economic element throughout the remainder of the region as well.

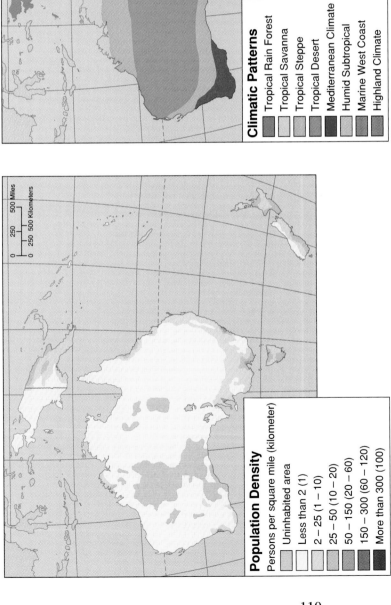

Climatic Patterns

Tropical Rain Forest
Tropical Savanna
Tropical Steppe
Tropical Desert
Mediterranean Climate
Humid Subtropical
Marine West Coast
Highland Climate

Population Density

Persons per square mile (kilometer)

Uninhabited area
Less than 2 (1)
2 – 25 (1 – 10)
25 – 50 (10 – 20)
50 – 150 (20 – 60)
150 – 300 (60 – 120)
More than 300 (100)

Map 79b Population Density

The region's small population is remarkably diverse. To the north are the Melanesian New Guinea peoples while Europeans dominate the populations of Australia and New Zealand, both of which have significant indigenous populations. Throughout most of the smaller islands groups, the bulk of the population is Melanesian but with a scattering of Europeans. The distribution of population is extremely uneven with New Guinea, southeastern Australia, and New Zealand supporting the bulk of the region's people while the remainder of the region—meaning nearly all of Australia—is either sparsely populated or devoid of population altogether. Nowhere do population densities reach the levels they do in other major regions of the world and densities of 50 persons per square mile are the highest to be found. Population location is dependent on precipitation and population growth patterns are culturally variable.

Map 79c Climatic Patterns

Because of its nearly-uniform surface, with only a few low and scattered uplands, Australian climate is a consequence only of the two great climate controls—latitudinal position and location relative to continental margin and interior. The continent bestrides the 30th parallel of latitude and its climatic pattern is dominated by the subtropical high pressure system with dry air masses that are responsible for the existence of great deserts. Toward the equator, the desert grades into steppe, savanna, and tropical forest as the subtropical high gives way to equatorial low pressure and abundant precipitation. Toward the pole, arid land fades into more well-watered steppe grasslands, the Mediterranean type climate of the southern margins of the continent, and the marine west coast climate of the Australian southeast and New Zealand. This latter climate is where most of the region's people live.

Part VIII

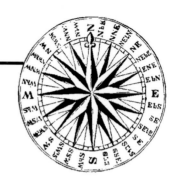

World Countries: Data Tables

Table A

World Countries: Area, Population, and Population Density, 1998

COUNTRY	AREA		POPULATION	DENSITY	
	(Mi^2)	(Km^2)	(1997)[a]	(Pop/Mi^2)	(Pop/Km^2)
Afghanistan	251,826	652,229	23,738,085	94	36
Albania	11,100	28,749	3,293,252	297	115
Algeria	919,595	2,381,750	29,830,370	32	13
Andorra	175	453	74,839	428	165
Angola	481,354	1,246,706	10,523,994	22	8
Antigua and Barbuda	171	443	68,175	399	154
Argentina	1,073,400	2,780,104	35,797,536	33	13
Armenia	11,506	29,801	3,465,611	301	116
Australia	2,966,155	7,682,337	18,438,824	6	2
Austria	32,377	83,856	8,054,078	249	96
Azerbaijan	33,436	86,599	7,735,918	231	89
Bahamas	5,382	13,939	262,034	49	19
Bahrain	267	692	603,318	2,260	872
Bangladesh	55,598	143,999	125,340,261	2,254	870
Barbados	166	430	257,731	1,553	599
Belarus	80,155	207,601	10,439,916	130	50
Belgium	11,783	30,518	10,203,563	866	334
Belize	8,866	22,963	224,663	25	10
Benin	43,475	112,600	5,902,178	136	52
Bhutan	18,200	47,138	1,865,191	102	40
Bolivia	424,165	1,098,587	7,669,868	18	7
Bosnia-Herzegovina	19,741	51,129	2,607,734	132	51
Botswana	231,800	600,362	1,500,765	6	2
Brazil	3,286,488	8,511,999	164,611,355	50	19
Brunei Darussalam	2,226	5,765	307,616	138	53
Bulgaria	42,823	110,912	8,652,745	202	78
Burkina Faso	105,869	274,201	10,891,159	103	40
Burundi	10,745	27,830	6,052,614	563	217
Cambodia	69,898	181,036	11,164,861	160	62
Cameroon	183,569	475,443	14,877,510	81	31
Canada	3,849,674	9,970,650	29,123,194	8	3
Cape Verde	1,557	4,033	393,843	253	98
Central African Republic	240,535	622,985	3,342,051	14	5
Chad	495,755	1,284,005	7,166,023	14	6
Chile	292,135	756,629	14,508,168	50	19
China	3,689,631	9,556,139	1,210,004,956	328	127
Colombia	440,831	1,141,752	37,418,290	85	33
Comoros	863	2,235	589,797	683	264
Congo (Zaire)	905,446	2,345,104	47,440,362	52	20
Congo Republic	132,047	342,002	2,583,198	20	8
Costa Rica	19,730	51,101	3,353,174	170	66
Cote d'Ivoire	124,518	322,501	14,966,218	120	46
Croatia	21,829	56,537	5,026,995	230	89
Cuba	42,804	110,862	10,999,041	257	99
Cyprus	3,593	9,306	752,808	210	81
Czech Republic	30,613	79,288	10,318,958	337	130
Denmark	16,638	43,092	5,268,775	317	122
Djibouti	8,958	23,201	434,116	48	19
Dominica	305	790	83,226	273	105
Dominican Republic	18,704	48,443	828,151	44	17
Ecuador	109,484	283,563	11,690,535	107	41
Egypt	386,662	1,001,454	64,791,891	168	65
El Salvador	8,124	21,041	5,661,827	697	269
Equatorial Guinea	10,831	28,052	442,516	41	16
Eritrea	45,300	117,327	3,589,687	79	31
Estonia	17,413	45,100	1,444,721	83	32
Ethiopia	483,123	1,251,288	58,732,577	122	47
Fiji	7,078	18,332	792,441	112	43
Finland	130,559	338,148	5,109,148	39	15
France	211,208	547,028	58,040,230	275	106
Gabon	103,347	267,669	1,190,159	12	4
Gambia	4,127	10,689	1,248,085	302	117
Georgia	26,911	69,699	5,174,642	192	74
Germany	137,882	357,114	84,068,216	610	235
Ghana	92,098	238,534	18,100,703	197	76
Greece	50,962	131,992	10,583,126	208	80
Grenada	133	344	95,537	718	277

(Continued on next page)

COUNTRY	AREA		POPULATION	DENSITY	
	(Mi^2)	(Km^2)	(1997)[a]	(Pop/Mi^2)	(Pop/Km^2)
Guatemala	42,042	108,889	11,558,407	275	106
Guinea	94,926	245,858	7,405,375	78	30
Guinea-Bissau	13,948	36,125	1,178,584	84	33
Guyana	83,000	214,970	706,116	9	3
Haiti	10,714	27,749	6,611,407	617	238
Honduras	43,277	112,087	5,751,384	133	51
Hungary	35,920	93,033	9,935,774	277	107
Iceland	36,769	95,232	272,550	7	3
India	1,237,062	3,203,989	967,612,804	782	302
Indonesia	742,410	1,922,841	209,774,138	283	109
Iran	632,457	1,638,063	67,540,002	107	41
Iraq	169,235	438,318	22,219,289	131	51
Ireland	27,137	70,285	3,555,500	131	51
Israel[b]	8,019	20,769	5,534,872	690	266
Italy	116,234	301,046	57,534,088	495	191
Jamaica	4,244	10,992	2,615,582	616	238
Japan	145,870	377,803	125,716,637	862	333
Jordan	35,135	91,000	4,324,638	123	48
Kazakhstan	1,049,156	2,717,313	16,895,572	16	6
Kenya	224,961	582,649	28,803,085	128	49
Kiribati	280	725	82,449	294	114
Korea, North	46,540	120,539	24,317,004	522	202
Korea, South	38,230	99,016	45,948,811	1,202	464
Kuwait	6,880	17,819	2,076,805	302	117
Kyrgyzstan	76,641	198,500	4,540,185	59	23
Laos	91,429	236,801	5,116,959	56	22
Latvia	24,595	63,701	2,437,649	99	38
Lebanon	4,015	10,399	3,858,736	961	371
Lesotho	11,720	30,355	2,007,814	171	66
Liberia	38,250	99,067	2,602,068	68	26
Libya	679,362	1,759,547	5,648,359	8	3
Liechtenstein	62	161	31,461	507	196
Lithuania	25,174	65,201	3,635,932	144	56
Luxembourg	998	2.585	422,474	423	163
Macedonia	9,928	25,714	2,113,866	213	82
Madagascar	226,658	587,044	14,061,627	62	24
Malawi	45,747	118,485	9,609,081	210	81
Malaysia	129,251	334,760	20,376,235	158	61
Maldives	115	298	280,391	2,438	941
Mali	478,767	1,240,006	9,945,383	21	8
Malta	122	316	379,365	3,110	1,201
Marshall Islands	70	181	60,652	866	335
Mauritania	395,956	1,025,525	2,411,317	6	2
Mauritius	788	2,041	1,154,272	1,465	566
Mexico	756,066	1,958,210	97,563,374	129	50
Micronesia	271	702	127,616	471	182
Moldova	13,012	33,701	4,475,232	344	133
Monaco	1	2	31,892	43,688	16,868
Mongolia	604,829	1,566,506	2,538,211	4	2
Morocco	275,114	712,545	30,391,423	110	43
Mozambique	308,642	799,382	18,165,476	59	23
Myanmar (Burma)	261,228	676,580	46,821,943	179	69
Namibia	317,818	823,148	1,727,183	5	2
Nauru	8	21	10,390	1,267	489
Nepal	56,827	147,182	22,641,061	398	154
Netherlands	16,133	41,784	15,653,091	970	375
New Zealand	103,519	268,114	3,358,275	32	13
Nicaragua	50,054	129,640	4,386,399	88	34
Niger	489,191	1,267,004	9,388,859	19	7
Nigeria	356,669	923,772	107,129,469	300	116
Norway	125,050	323,879	4,404,456	35	14
Oman	82,030	212,458	2,264,590	28	11
Pakistan	310,432	804,018	132,185,299	426	164
Palau	177	458	17,240	97	38
Panama	29,157	75,517	2,693,417	92	36
Papua New Guinea	178,704	462,843	4,496,221	25	10
Paraguay	157,048	406,754	5,651,634	36	14
Peru	496,225	1,285,222	24,949,512	50	19
Philippines	115,831	300,002	76,103,564	657	254
Poland	120,728	312,685	38,700,291	321	124
Portugal	35,516	91,986	9,867,654	278	107

(Continued on next page)

COUNTRY	AREA		POPULATION	DENSITY	
	(Mi^2)	(Km^2)	(1997)[a]	(Pop/Mi^2)	(Pop/Km^2)
Qatar	4,416	11,437	665,485	151	58
Romania	91,699	237,500	21,399,114	233	90
Russia	6,592,849	17,075,469	147,987,101	22	9
Rwanda	10,169	26,338	7,737,537	761	294
St. Kitts and Nevis	104	269	41,803	402	155
St. Lucia	238	616	159,639	671	259
St. Vincent/Grenadines	150	388	119,092	794	307
Samoa	1,093	2,831	219,509	201	78
San Marino	24	62	24,714	1,030	398
São Tomé and Principe	372	963	147,865	397	153
Saudi Arabia	830,000	2,149,699	20,087,965	24	9
Senegal	75,951	196,713	9,403,546	124	48
Seychelles	175	453	78,142	447	172
Sierra Leone	27,925	72,326	4,891,546	175	68
Singapore	246	637	3,461,929	14,073	5,434
Slovak Republic	18,933	49,036	5,393,016	285	110
Slovenia	7,819	20,251	1,945,998	0	0
Solomon Islands	10,954	28,371	426,855	39	15
Somalia	246,201	637,660	9,940,232	40	16
South Africa	433,680	1,123,231	42,327,458	98	38
Spain	194,885	504,752	39,244,195	201	78
Sri Lanka	24,962	64,652	18,762,075	752	290
Sudan	967,500	2,505,824	32,594,128	34	13
Suriname	63,251	163,820	443,446	7	3
Swaziland	6,704	17,363	1,031,600	154	59
Sweden	173,732	449,966	8,946,193	51	20
Switzerland	15,943	41,292	7,248,984	455	176
Syria	71,498	185,180	16,137,899	226	87
Taiwan	13,900	36,001	21,655,515	1,558	602
Tajikistan	55,251	143,100	6,013,855	109	42
Tanzania	364,900	945,090	29,460,753	81	31
Thailand	198,115	513,118	59,450,818	300	116
Togo	21,925	56,786	4,735,610	216	83
Tonga	290	751	107,335	370	143
Trinidad and Tobago	1,980	5,128	1,273,141	643	248
Tunisia	63,170	163,610	9,183,097	145	56
Turkey	300,948	779,455	63,528,225	211	82
Turkmenistan	188,456	488,101	4,225,351	22	9
Tuvalu	10	26	10,297	1,030	398
Uganda	93,104	241,139	20,604,874	221	85
Ukraine	233,090	603,703	50,684,635	217	84
United Arab Emirates	32,278	83,600	2,262,309	70	27
United Kingdom	94,248	244,102	58,610,182	622	240
United States	3,787,425	9,809,425	267,954,767	71	27
Uruguay	68,500	177,415	3,261,707	48	18
Uzbekistan	172,742	447,402	23,860,452	138	53
Vanatu	4,707	12,191	181,358	39	15
Venezuela	352,145	912,055	22,396,407	64	25
Vietnam	128,066	331,691	75,123,880	587	226
Yemen	205,356	531,872	13,972,477	68	26
Yugoslavia (Serbia-Montenegro)	26,913	69,705	10,655,317	396	153
Zambia	290,586	752,617	9,349,975	32	12
Zimbabwe	150,873	390,761	11,423,175	76	29

[a]Primary source for population figures: United Nations Population Division and International Labour Organisation.

[b]The figures for Israel do not include the West Bank and Gaza. The former has an area of approximately 2,300 square miles, a population of 1.8 million, and a density of about 780 persons per square mile; the latter has an area of 100 square miles, a population of nearly 900,000 persons, and an estimated population density of 8,700 persons per square mile.

Sources: *The World Almanac and Book of Facts, 1998* (St. Martin's Press, New York, 1998); *The New York Times 1998 Almanac* (Penguin Putnam, Inc., New York, 1998); U.S. Census Bureau, *World Population Profile* (1998).

Table B

World Countries: Form of Government, Capital City, Major Languages

Notes: Unless indicated otherwise, republics are multi-party. "Theocratic" normally refers to fundamentalist Islamic rule. "Transitional" governments are those still in the process of change from a previous form (eg. single-party communist state to multi-party republic).

COUNTRY	GOVERNMENT	CAPITAL	MAJOR LANGUAGES
Afghanistan	Theocratic republic/transitional	Kabul	Dari, Pashton, Uzbek, Turkmen
Albania	Democratic	Tiranë	Albanian, Greek
Algeria	Military-transitional	Algiers	Arabic, Berber dialects, French
Andorra	Parliamentary democracy	Andorra	Catalán, French, Spanish
Angola	Multi-party democracy	Luanda	Portugese, indigenous
Antigua and Barbuda	Parliamentary democracy	St. John's	English, local dialects
Argentina	Federal republic	Buenos Aires	Spanish, English, Italian, German
Armenia	Republic	Yerevan	Armenian, Azerbaijani, Russian
Australia	Federal parliamentary	Canberra	English, indigenous
Austria	Federal republic	Vienna	German
Azerbaijan	Republic	Baku	Azerbaijani, Russian, Armenian
Bahamas	Independent commonwealth	Nassau	English
Bahrain	Traditional monarchy	Al Manamah	Arabic, English, Farsi, Urdu
Bangladesh	Republic	Dhaka	Bangla, English
Barbados	Parliamentary state	Bridgetown	English
Belarus	Republic	Minsk	Byelorussian, Russian
Belgium	Constitutional monarchy	Brussels	Dutch (Flemish), French, German
Belize	Parliamentary state	Belmopan	English, Spanish, Garifuna, Mayan
Benin	Multi-party republic	Porto-Novo	French, Fon, Adja, indigenous
Bhutan	Monarchy	Thimphu	Dzongha, Tibetan, Nepalese
Bolivia	Republic	La Paz	Spanish, Quechua, Aymara
Bosnia-Herzegovina	Republic/transitional	Sarajevo	Serb, Croat, Albanian
Botswana	Parliamentary republic	Gaborone	English, Tswania
Brazil	Federal republic	Brasília	Portugese, Spanish, English, French
Brunei	Constitutional monarchy	Bandar	Malay, English, Chinese
Bulgaria	Republic/transitional	Sofia	Bulgarian
Burkina Faso	Provisional military	Ouagadougou	French, indigenous
Burundi	Republic/transitional	Bujumbura	French, Kirundi, Swahili
Cambodia	Transitional UN management	Phnom Penh	Khmer, French
Cameroon	Multi-party republic	Yaoundé	English, French, indigenous
Canada	Federal parliamentary	Ottawa	English, French
Cape Verde	Republic	Praia	Portugese, Crioulu
Central African Republic	Republic	Bangui	French, Sango, Arabic, indigenous
Chad	Republic	N'Djamena	Arabic, French, indigenous
Chile	Republic	Santiago	Spanish
China	Single-party communist state	Beijing	Chinese dialects
Colombia	Republic	Bogotá	Spanish
Comoros	Republic	Moroni	Arabic, French, Shaafi Islam
Congo (Zaire)	Republic	Kinshasa	French, Lingala, Kingwana, Kikongo, Tshuliba
Congo Republic	Multi-party republic	Brazzaville	French, indigenous
Costa Rica	Republic	San José	Spanish
Cote d'Ivoire	Republic	Abidjan	French, indigenous
Croatia	Parliamentary democracy	Zaghreb	Croat, Serb
Cuba	Single-party communist state	Havana	Spanish
Cyprus	Republic	Nicosia	Greek, Turkish, English
Czech Republic	Federal republic	Prague	Czech, Slovak, Hungarian
Denmark	Constitutional monarchy	Copenhagen	Danish
Djibouti	Republic	Djibouti	French, Somali, Afar, Arabic
Dominica	Republic	Rosseau	English, French
Dominican Republic	Republic	Santo Domingo	Spanish
Ecuador	Republic	Quito	Spanish, Quechua, indigenous
Egypt	Republic	Cairo	Arabic
El Salvador	Republic	San Salvador	Spanish, Nahua
Equatorial Guinea	Republic	Malabo	Spanish, indigenous, English
Eritrea	Transitional government	Asmara	Afar, Bilen, Hadareb, Kunuama, Nara
Estonia	Republic	Tallinn	Estonian, Russian
Ethiopia	Republic/transitional	Addis Ababa	Amharic, Tigrinya, Orominga, Arabic
Fiji	Republic	Suva	English, Fijian, Hindustani
Finland	Republic	Helsinki	Finnish, Swedish
France	Republic	Paris	French
Gabon	Single-party republic	Libreville	French, Fang, indigenous
Gambia	Republic	Banjul	English, Malinke, Wolof, Fula
Georgia	Republic	Tbilisi	Georgian, Russian, Armenian
Germany	Federal republic	Berlin and Bonn	German
Ghana	Parliamentary democracy	Accra	English, Akan, indigenous
Greece	Parliamentary democracy	Athens	Greek

(Continued on next page)

COUNTRY	GOVERNMENT	CAPITAL	MAJOR LANGUAGES
Grenada	Parliamentary state	St. George's	English, French
Guatemala	Republic	Guatemala	Spanish, indigenous
Guinea	Republic/transitional	Conakry	French, indigenous
Guinea-Bissau	Single-party republic	Bissau	Portugese, Crioulo, indigenous
Guyana	Republic	Georgetown	English, indigenous
Haiti	Republic	Port-au-Prince	Creole, French
Honduras	Republic	Tegucigalpa	Spanish, indigenous
Hungary	Republic	Budapest	Hungarian
Iceland	Republic	Reykjavík	Icelandic
India	Federal republic	New Delhi	English, Hindi, Telugu, Bengali
Indonesia	Republic	Jakarta	Indonesian, Javanese, Sundanese
Iran	Theocratic republic	Tehran	Farsi, Turkish, Kurdish, Arabic, English
Iraq	Single-party republic	Baghdad	Arabic, Kurdish, Assyrian, Armenian
Ireland	Republic	Dublin	English, Irish Gaelic
Israel	Republic	Jerusalem	Hebrew, Arabic, Yiddish
Italy	Republic	Rome	Italian
Jamaica	Parliamentary state	Kingston	English, Creole
Japan	Constitutional monarchy	Tokyo	Japanese
Jordan	Constitutional monarchy	Amman	Arabic
Kazakhstan	Republic/transitional	Alma-Ata	Kazakh, Russian, German, Ukranian
Kenya	Republic	Nairobi	English, Swahili, indigenous
Kiribati	Republic	Tarawa	English, Gilbertese
Korea, North	Single-party communist state	Pyongyang	Korean
Korea, South	Republic	Seoul	Korean
Kuwait	Constitutional monarchy	Kuwait	Arabic, English
Kyrgyzstan	Republic/transitional	Bishkek	Kirghiz, Russian, Uzbek
Laos	Single-party communist state	Viangchan	Lao, French, Thai, indigenous
Latvia	Republic	Riga	Latvian, Russian
Lebanon	Republic	Beirut	Arabic, French, Armenian, English
Lesotho	Constitutional monarchy	Maseru	English, Sesotho, Zulu, Xhosa
Liberia	Republic	Monrovia	English, indigenous
Libya	Single-party parliamentary state	Tripoli	Arabic
Liechtenstein	Constitutional monarchy	Vaduz	German
Lithuania	Republic	Vilnius	Lithuanian, Russian, Polish
Luxembourg	Constitutional monarchy	Luxembourg	French, Luxembourgish, German
Macedonia	Republic/transitional	Skopje	Macedonian, Albanian
Madagascar	Military/transitional	Antananarivo	Malagasy, French
Malawi	Single-party republic	Lilongwe	Chichewa, English, Tombuka
Malaysia	Constitutional monarchy	Kuala Lumpur	Malay, Chinese, English
Maldives	Republic	Male	Divehi
Mali	Single-party republic	Bamako	French, Bambara, indigenous
Malta	Parliamentary democracy	Valletta	English, Maltese
Marshall Islands	Constitutional government	Majuro	English, Polynesian dialects, Japanese
Mauritania	Republic/transitional	Nouakchott	Arabic, French, indigenous
Mauritius	Parliamentary state	Port Louis	English, Creole, Bhojpun, Hindi
Mexico	Federal republic	Mexico City	Spanish, indigenous
Micronesia	Republic	Kolonia	English, Malay-Polynesian languages
Moldova	Republic	Kishinev	Moldavian, Russian, Ukrainian
Monaco	Constitutional monarchy	Monaco	French, English, Italian, Monegasque
Mongolia	Single-party communist state	Ulan Bator	Khalkha Mongol, Kazakh, Russian
Morocco	Constitutional monarchy	Rabat	Arabic, Berber dialects, French
Mozambique	Republic	Maputo	Portuguese, indigenous
Myanmar (Burma)	Provisional military	Rangoon	Burmese, indigenous
Namibia	Republic	Windhoek	Afrikaans, English, German, indigenous
Nauru	Republic	Yaren district	Nauruan, English
Nepal	Constitutional monarchy	Kathmandu	Nepali, Maithali, Bhojpuri, indigenous
Netherlands	Constitutional monarchy	Amsterdam	Dutch
New Zealand	Parliamentary state	Wellington	English, Maori
Nicaragua	Republic	Managua	Spanish, English, indigenous
Niger	Provisional military	Niamey	French, Hausa, Djerma, indigenous
Nigeria	Provisional military	Lagos/Abuja	English, Hausa, Fulani, Yorbua, Ibo
Norway	Constitutional monarchy	Oslo	Norwegian, Lapp
Oman	Traditional monarchy	Muscat	Arabic, English, Baluchi, Urdu
Pakistan	Federal Islamic republic	Islamabad	English, Urdu, Punjabi, Pashto, Sindhi
Palau	Republic	Koror state	English, Palauan
Panama	Republic	Panama	Spanish, English, indigenous
Papua New Guinea	Parliamentary state	Port Moresby	English, Motu, Pidgin, indigenous
Paraguay	Republic (authoritarian)	Asunción	Spanish, Guarani
Peru	Republic	Lima	Quechua, Spanish, Aymara
Philippines	Republic	Manila	English, Philipino, Tagalog
Poland	Republic	Warsaw	Polish
Portugal	Republic	Lisbon	Portuguese
Qatar	Traditional monarchy	Doha	Arabic, English

(Continued on next page)

COUNTRY	GOVERNMENT	CAPITAL	MAJOR LANGUAGES
Romania	Republic	Bucharest	Romanian, Hungarian, German
Russia	Republic	Moscow	Russian, Tatar, Ukranian
Rwanda	Provisional military	Kigali	French, Kinyarwanda
St. Kitts and Nevis	Parliamentary state	Basseterre	English
St. Lucia	Parliamentary state	Castries	English, French
St. Vincent/Grenadines	Parliamentary state	Kingstown	English, French
Samoa	Constitutional monarchy	Apia	Samoan, English
San Marino	Republic	San Marino	Italian
São Tomé and Príncipe	Republic	São Tomé	Portuguese, Fang
Saudi Arabia	Islamic monarchy	Riyadh	Arabic
Senegal	Republic	Dakar	French, Wolof, indigenous
Seychelles	Republic	Victoria	English, French, Creole
Sierra Leone	Republic	Freetown	English, Krio, indigenous
Singapore	Republic	Singapore	Mandarin Chinese, English, Malay
Slovak Republic	Parliamentary democracy	Bratislava	Slovak, Hungarian, Polish
Slovenia	Republic/transitional	Ljubljana	Slovene
Solomon Islands	Parliamentary state	Honiara	English, Malay-Polynesian languages
Somalia	Provisional military	Mogadishu	Arabic, Somali, English, Italian
South Africa	Republic	Pretoria	Afrikaans, English, Zulu, Xhosa, other
Spain	Parliamentary monarchy	Madrid	Spanish, Catalan, Galician, Basque
Sri Lanka	Republic	Colombo	English, Sinhala, Tamil
Sudan	Islamic republic	Khartoum	Arabic, Dinka, Nubian
Suriname	Military/transitional	Paramaribo	Dutch, Sranan Tongo, English
Swaziland	Monarchy	Mbabane	English, Swahili
Sweden	Constitutional monarchy	Stockholm	Swedish
Switzerland	Federal republic	Bern	German, French, Italian, Romansch
Syria	Republic (authoritarian)	Damascus	Arabic, Kurdish, Armenian, Aramaic
Taiwan	Single-party republic	T'aipei	Chinese dialects
Tajikistan	Republic	Dushanbe	Tajik, Uzbek, Russian
Tanzania	Republic	Dar es Salaam	English, Swahili, indigenous
Thailand	Constitutional monarchy	Bangkok	Thai, indigenous
Togo	Single-party republic	Lomé	French, indigenous
Tonga	Constitutional monarchy	Nuku'alofa	Tongan, English
Trinidad and Tobago	Parliamentary democracy	Port of Spain	English, Hindi, French, Spanish
Tunisia	Republic	Tunis	Arabic, French
Turkey	Parliamentary democracy	Ankara	Turkish, Kurdish, Arabic
Turkmenistan	Republic	Ashkhabad	Turkmen, Russian, Uzbek, Kazakh
Tuvalu	Parliamentary democracy	Funafuti	Tuvaluan, English
Uganda	Republic	Kampala	English, Luganda, Swahili, indigenous
Ukraine	Republic	Kiev	Ukranian, Russian
United Arab Emirates	Federated monarchy	Abu Dhabi	Arabic, English, Farsi, Hindi, Urdu
United Kingdom	Constitutional monarchy	London	English, Welsh, Gaelic
United States	Federal republic	Washington	English, Spanish
Uruguay	Republic	Montevideo	Spanish
Uzbekistan	Republic	Tashkent	Uzbek, Russian, Kazakh, Tajik, Tatar
Vanatu	Republic	Port Vila	English, French, Bislama (pidgin)
Venezuela	Federal republic	Caracas	Spanish, indigenous
Vietnam	Single-party communist state	Hanoi	Vietnamese, French, Chinese, English
Yemen	Republic	San'a	Arabic
Yugoslavia	Republic	Belgrade	Serb, Albanian, Hungarian
Zambia	Single-party republic	Lusaka	English, Tonga, Lozi, other indigenous
Zimbabwe	Republic	Harare	English, ChiShona, SiNdebele

Sources: *The World Almanac and Book of Facts, 1998* (St. Martin's Press, New York, 1998); *The New York Times 1998 Almanac* (Penguin Putnam, Inc., New York, 1998); U.S. Census Bureau, *World Population Profile* (1997).

Table C
World Countries: Basic Economic Indicators

COUNTRY	GROSS NATIONAL PRODUCT (GNP) 1995 Total (Millions $US)	Per Capita ($US)	GROSS DOMESTIC PRODUCT (GDP) 1995 Total (Millions $US)	Per Capita ($US)	PURCHASING POWER PARITY (PPP) 1995 Total (Millions $US)	Per Capita ($US)	AVERAGE ANNUAL GROWTH RATE GDP 1975–1995 (percent) 1975–85	1985–95	DISTRIBUTION OF GDP, 1995 (percent) Agriculture	Industry	Services
AFRICA											
Algeria	44,609	1,600	41,435	1,474	157,410	5,600	6.1	0.3	13	47	41
Angola	4,422	410	3,722	344	12,655	1,170	X	0.4	12	59	28
Benin	2,034	370	1,522	289	8,730	1,660	3.7	X	34	12	53
Botswana	4,381	3,010	4,318	2,978	8,164	5,630	10.8	7.1	5	46	48
Burkina Faso	2,417	230	2,325	222	8,278	790	4.5	2.9	34	27	39
Burundi	984	160	1,062	175	3,881	640	4.6	0.9	56	18	26
Cameroon	8,615	650	7,931	601	30,342	2,300	8.0	(2.1)	39	23	38
Central African Republic	1,123	340	1,128	345	3,535	1,080	1.4	1.0	44	13	43
Chad	1,144	180	1,138	180	4,498	710	2.4	2.6	44	22	35
Congo (Zaire)	5,313	120	8,770	243	17,056	473	X	X	30	33	36
Congo Republic	1,784	680	2,163	834	6,431	2,480	8.4	(0.3)	10	38	51
Cote d'Ivoire	9,248	660	10,069	735	24,238	1,770	3.5	0.1	31	20	50
Egypt	45,507	790	47,349	763	241,553	3,890	9.5	2.2	20	21	59
Equatorial Guinea	152	380	169	421	X	X	X	X	50	33	17
Eritrea	X	X	X	X	X	X	X	X	11	20	69
Ethiopia	5,722	100	5,287	94	25,946	460	X	3.6	57	10	33
Gabon	3,759	3,490	4,691	4,360	3,943	3,983	1.9	(0.5)	8	52	40
Gambia	354	320	384	346	1,055	950	3.4	2.6	28	15	58
Ghana	6,719	390	6,315	364	35,196	2,030	(0.9)	4.4	46	16	38
Guinea	3,593	550	3,686	502	3,694	577	X	X	24	31	45
Guinea-Bissau	265	250	257	240	855	800	2.6	3.7	46	24	30
Kenya	7,583	280	9,095	335	38,825	1,430	4.5	3.5	29	17	54
Lesotho	1,519	770	1,029	508	2,513	1,240	4.0	7.0	11	40	49
Liberia	2,300	770	1,202	467	X	X	0.2	X	X	X	X
Libya	32,900	6,510	21,864	4,984	X	X	1.0	X	X	X	X
Madagascar	3,178	230	3,198	215	10,114	680	0.2	0.9	34	13	53
Malawi	1,623	170	1,465	151	7,448	770	4.0	2.1	42	27	31
Mali	2,410	250	2,431	224	6,045	560	2.6	4.3	46	17	37
Mauritania	1,049	460	1,068	470	3,684	1,620	1.1	3.1	27	30	43
Mauritius	3,815	3,380	3,919	3,508	14,823	13,270	X	5.9	9	33	58
Morocco	29,545	1,110	32,412	1,222	92,038	3,470	5.4	2.6	14	33	53
Mozambique	1,353	80	1,469	85	15,707	910	X	5.8	33	12	55
Namibia	3,098	2,000	3,033	1,974	6,298	4,100	X	3.2	14	29	56
Niger	1,961	220	1,860	203	6,863	750	0.8	1.4	39	18	4
Nigeria	28,411	260	40,477	362	146,355	1,310	0.4	3.8	28	53	18
Rwanda	1,128	180	1,128	218	2,799	540	6.1	(4.8)	37	17	46
Senegal	5,070	600	4,867	586	15,211	1,830	2.9	2.8	20	18	62
Sierra Leone	762	180	824	196	2,601	620	2.5	(1.5)	42	27	31
Somalia	3,300	500	917	106	7,923	933	6.2	X	X	X	X
South Africa	130,918	3,160	136,035	3,281	217,277	5,240	2.3	1.1	5	31	64
Sudan	26,600	860	5,989	239	17,852	725	3.6	X	X	X	X
Swaziland	1,051	1,170	1,073	1,252	2,528	2,950	3.8	3.5	X	X	X
Tanzania	3,703	120	3,602	120	20,117	670	X	3.6	58	17	24
Togo	1,266	310	1,263	309	4,739	1,160	2.5	0.9	38	21	41
Tunisia	16,369	1,820	18,035	2,007	47,272	5,260	5.9	3.3	12	20	59
Uganda	4,668	240	5,655	287	29,337	1,490	X	5.7	50	14	36

(Continued on next page)

COUNTRY	GROSS NATIONAL PRODUCT (GNP) 1995 Total (Millions $US)	Per Capita ($US)	GROSS DOMESTIC PRODUCT (GDP) 1995 Total (Millions $US)	Per Capita ($US)	PURCHASING POWER PARITY (PPP) 1995 Total (Millions $US)	Per Capita ($US)	AVERAGE ANNUAL GROWTH RATE GDP 1975–1995 (percent) 1975–85	1985–95	DISTRIBUTION OF GDP, 1995 (percent) Agriculture	Industry	Services
Zambia	3,605	400	4,073	504	8,000	990	0.1	0.5	22	40	37
Zimbabwe	5,933	540	6,522	583	23,947	2,140	2.6	2.1	15	36	48

NORTH AND CENTRAL AMERICA

COUNTRY	Total (Millions $US)	Per Capita ($US)	Total (Millions $US)	Per Capita ($US)	Total (Millions $US)	Per Capita ($US)	1975–85	1985–95	Agriculture	Industry	Services
Belize	560	2,630	578	2,714	1,197	5,620	4.0	6.5	20	28	53
Canada	573,695	19,380	568,928	19,350	643,904	21,900	3.6	2.2	3	31	66
Costa Rica	8,884	2,610	9,233	2,696	20,270	5,920	2.9	4.5	17	24	58
Cuba	14,700	1,300	X	X	X	X	X	X	X	X	X
Dominican Republic	11,390	1,460	11,277	1,442	29,258	3,740	3.6	3.6	15	22	64
El Salvador	9,057	1,610	9,471	1,673	14,721	2,600	(1.1)	4.0	14	22	65
Guatemala	14,255	1,340	14,489	1,364	35,943	3,290	2.4	3.3	25	19	56
Haiti	1,777	250	2,043	287	6,554	920	2.0	(2.8)	44	12	44
Honduras	3,566	600	3,927	696	11,534	2,040	4.6	3.2	21	33	46
Jamaica	3,803	1,510	4,406	1,785	12,167	4,930	(1.4)	3.6	9	38	53
Mexico	304,596	3.320	250,038	2,743	615,229	6,750	4.7	1.0	8	26	67
Nicaragua	1,659	380	1,911	463	10,019	2,430	X	X	33	20	46
Panama	7,235	2,750	7,413	2,818	16,470	6,260	4.6	2.3	11	15	74
Trinidad and Tobago	4,851	3,770	5,327	4,139	12,368	9,610	3.2	(0.5)	3	42	54
United States	7,100,007	26,980	6,952,020	26,026	7,206,763	26,980	2.7	2.5	2	26	72

SOUTH AMERICA

COUNTRY	Total (Millions $US)	Per Capita ($US)	Total (Millions $US)	Per Capita ($US)	Total (Millions $US)	Per Capita ($US)	1975–85	1985–95	Agriculture	Industry	Services
Argentina	278,431	8,030	281,060	8,084	203,790	8,450	0.2	2.6	6	31	63
Bolivia	5,905	800	6,131	827	19,870	2,680	2.2	3.0	33	32	35
Brazil	579,787	3,640	688,085	4,327	874,583	5,500	X	X	14	37	49
Chile	59,151	4,160	67,297	4,736	138,263	9,730	2.7	6.7	8	41	50
Colombia	70,263	1,910	76,112	2,125	226,703	6,330	4.0	4.4	14	32	54
Ecuador	15,997	1,390	17,939	1,565	52,258	4,560	4.8	2.7	12	36	52
Guyana	493	590	595	717	2,141	2,580	(1.7)	2.4	36	37	27
Paraguay	8,158	1,690	7,743	1,604	17,526	3,630	6.6	3.5	24	22	54
Peru	55,019	2,310	57,424	2,440	89,422	3,800	1.4	2.0	7	38	55
Suriname	360	880	335	784	905	2,120	3.6	3.7	26	26	48
Uruguay	16,458	5,170	17,847	5,602	21,346	6,700	0.9	3.5	9	26	65
Venezuela	65,382	3,020	75,016	3,434	176,936	8,100	1.0	2.8	5	38	56

ASIA

COUNTRY	Total (Millions $US)	Per Capita ($US)	Total (Millions $US)	Per Capita ($US)	Total (Millions $US)	Per Capita ($US)	1975–85	1985–95	Agriculture	Industry	Services
Afghanistan	12,800	600	X	X	X	X	0.0	0.0	X	X	X
Armenia	2,752	730	2,843	783	8,281	2,280	7.3	(10.3)	44	35	20
Azerbaijan	3,601	480	3,473	461	10,995	1,460	X	X	27	32	41
Bangladesh	28,599	240	29,110	246	163,156	1,380	4.4	4.0	31	18	52
Bhutan	295	420	304	172	2,283	1,290	X	6.3	40	32	28
Cambodia	2,718	270	2,771	276	X	X	X	X	51	14	34
China	744,890	620	697,647	572	3,624,06	2,970	8.7	9.6	21	48	31
Georgia	2,358	440	2,325	427	8,066	1,480	6.0	(16.7)	67	22	11
India	319,660	340	324,082	349	1,319,18	1,420	5.0	5.2	29	29	41
Indonesia	190,105	980	109,079	1,003	783,916	3,970	7.2	7.2	17	42	41
Iran	X	X	110,771	1,756	379,426	5,550	1.0	1.9	25	34	40
Iraq	X	X	48,422	2,755	X	X	0.6	X	X	X	X
Israel	87,875	15,920	91,965	16,645	92,268	16,700	3.7	5.3	X	X	X
Japan	4,963,587	39,640	5,108,540	40,846	2,743,741	21,930	4.1	2.9	2	38	60

(Continued on next page)

COUNTRY	GROSS NATIONAL PRODUCT (GNP) 1995 Total (Millions $US)	GROSS NATIONAL PRODUCT (GNP) 1995 Per Capita ($US)	GROSS DOMESTIC PRODUCT (GDP) 1995 Total (Millions $US)	GROSS DOMESTIC PRODUCT (GDP) 1995 Per Capita ($US)	PURCHASING POWER PARITY (PPP) 1995 Total (Millions $US)	PURCHASING POWER PARITY (PPP) 1995 Per Capita ($US)	AVERAGE ANNUAL GROWTH RATE GDP 1975–1995 (percent) 1975–85	AVERAGE ANNUAL GROWTH RATE GDP 1975–1995 (percent) 1985–95	DISTRIBUTION OF GDP, 1995 (percent) Agriculture	DISTRIBUTION OF GDP, 1995 (percent) Industry	DISTRIBUTION OF GDP, 1995 (percent) Services
Jordan	6,354	1,510	6,105	1,187	21,292	4,140	X	X	8	27	65
Kazakhstan	22,143	1,330	21,413	1,273	51,124	3,040	X	X	12	30	57
Korea, North	X	X	X	X	X	X	X	X	X	X	X
Korea, South	453,137	9,700	455,476	10,142	518,699	11,550	8.6	8.4	7	43	50
Kuwait	28,941	17,390	26,650	15,760	37,303	22,060	(3.2)	3.7	0	53	46
Kyrgyzstan	3,158	700	3,054	685	8,028	1,800	X	X	44	24	32
Laos	1,694	350	1,760	361	7,404	1,709	X	5.2	52	18	30
Lebanon	10,673	2,660	11,143	3,703	X	X	X	X	7	24	69
Malaysia	78,321	3,890	85,311	4,236	191,733	9,520	6.7	7.4	13	43	44
Mongolia	767	310	861	349	4,951	2,010	X	0.3	17	26	57
Myanmar (Burma)	X	X	X	X	28,277	696	5.8	X	63	9	28
Nepal	4,391	200	4,232	197	24,460	1,140	3.7	4.7	42	22	36
Oman	10,578	4,820	12,102	5,483	20,635	9,350	11.3	4.3	3	53	43
Pakistan	59,991	460	60,649	445	301,128	2,210	6.7	5.1	26	24	50
Philippines	71,865	1,050	74,180	1,093	187,236	2,760	2.8	3.4	22	32	46
Saudi Arabia	133,540	7,040	125,501	6,875	158,636	8,690	1.6	2.9	6	50	43
Singapore	79,831	26,730	83,695	25,156	75,223	22,610	7.4	7.9	0	36	64
Sri Lanka	12,616	700	12,915	720	58,983	3,290	5.9	3.8	23	25	52
Syria	15,780	1,120	16,783	1,182	80,247	5,650	6.5	4.2	29	24	48
Tajikistan	1,976	340	1,999	343	5,653	970	X	X	27	34	39
Thailand	159,630	2,470	167,056	2,868	449,046	7,710	6.9	9.0	11	40	49
Turkey	169,452	2,780	164,789	2,709	335,217	5,510	4.4	4.3	16	31	53
Turkmenistan	4,125	920	3,917	961	15,469	4,217	X	X	32	30	38
United Arab Emirates	42,806	17,400	39,107	17,696	31,912	14,440	6.8	X	2	57	40
Uzbekistan	21,979	970	21,556	947	53,946	2,370	X	X	33	34	34
Vietnam	17,634	240	20,351	276	X	X	X	6.3	28	30	42
Yemen	4,044	260	4,790	319	X	X	X	X	22	27	51
EUROPE											
Albania	2,199	670	2,192	648	X	X	X	3.3	56	21	23
Austria	216,547	26,890	233,427	29,015	171,519	21,320	2.2	2.5	2	34	63
Belarus	21,356	2,070	20,561	1,986	43,996	4,250	X	X	13	35	52
Belgium	250,710	24,710	269,081	26,571	218,338	21,560	1.8	2.1	2	30	68
Bosnia-Herzegovina	X	X	X	X	X	X	X	X	X	X	X
Bulgaria	11,225	1,330	12,366	1,453	39,992	4,700	X	(1.6)	13	34	53
Croatia	15,508	3,250	18,081	4,014	X	X	X	X	12	25	62
Czech Republic	39,990	3,870	44,772	4,362	100,270	9,770	X	(0.8)	6	39	55
Denmark	156,027	29,890	172,220	32,973	114,854	21,990	2.5	1.7	4	29	67
Estonia	6,136	2,860	60,776	40,838	6,279	4,220	X	(4.2)	8	28	64
Finland	105,174	20,580	125,432	24,561	94,684	18,540	2.8	1.3	6	37	57
France	1,451,051	24,990	1,536,089	26,437	1,230,643	21,280	2.3	2.1	2	27	71
Germany	2,252,343	27,510	2,415,764	29,607	1,641,671	20,120	X	X	1	38	61
Greece	85,885	8,210	90,550	8,662	121,685	11,640	2.9	1.3	21	36	43
Hungary	41,129	4,120	43,712	4,325	67,508	6,680	3.1	(0.1)	8	33	59
Iceland	6,686	24,950	7,052	26,215	5,671	21,080	4.2	2.0	12	28	60
Ireland	52,765	14,710	60,780	17,141	63,119	17,800	4.0	4.5	8	10	82
Italy	1,088,085	19,020	1,086,932	19,001	1,154,377	20,180	2.8	2.1	3	31	60
Latvia	5,708	2,270	5,689	2,243	8,521	3,360	4.1	(5.2)	9	31	60
Lithuania	7,070	1,900	7,089	1,897	15,392	4,120	X	X	11	36	53
Macedonia	1,813	860	1,975	937	X	X	X	X	X	X	X

(Continued on next page)

| COUNTRY | GROSS NATIONAL PRODUCT (GNP) 1995 | | GROSS DOMESTIC PRODUCT (GDP) 1995 | | PURCHASING POWER PARITY (PPP) 1995 | | AVERAGE ANNUAL GROWTH RATE GDP 1975–1995 (percent) | | DISTRIBUTION OF GDP, 1995 (percent) | | |
	Total (Millions $US)	Per Capita ($US)	Total (Millions $US)	Per Capita ($US)	Total (Millions $US)	Per Capita ($US)	1975–85	1985–95	Agriculture	Industry	Services
Moldova	3,996	920	3,518	793	17,434	3,976	X	X	50	23	22
Netherlands	371,039	24,000	395,900	25,572	307,782	19,880	1.9	2.5	3	27	70
Norway	136,077	31,250	145,954	33,692	97,253	22,450	4.3	2.5	3	35	62
Poland	107,839	2,790	117,663	3,052	209,635	5,430	X	0.8	6	39	54
Portugal	96,689	9,740	102,337	10,427	124,552	12,690	2.5	3.0	6	38	56
Romania	33,488	1,480	35,533	1,563	99,776	4,390	X	(2.4)	21	40	39
Russian Federation	331,948	2,240	346,383	2,333	715,577	4,820	4.4	(3.9)	7	38	55
Slovak Republic	15,848	2,950	17,414	3,262	19,217	3,600	X	(0.8)	6	33	61
Slovenia	16,328	8,200	18,550	9,636	X	X	X	X	5	39	57
Spain	532,347	13,580	558,617	14,097	585,687	14,780	1.7	2.9	3	33	64
Sweden	209,720	23,750	228,679	26,022	169,696	19,310	1.8	1.2	2	32	66
Switzerland	286,014	40,630	300,508	41,935	178,433	24,900	0.8	1.4	X	X	X
Ukraine	84,084	1,630	80,127	1,548	126,287	2,440	X	X	18	42	40
United Kingdom	1,094,734	18,700	1,105,822	19,040	1,120,925	19,300	1.8	2.2	2	32	66
Yugoslavia (Serbia-Montenegro)	X	X	X	X	55,523	5,467	X	X	X	X	X
OCEANIA											
Australia	337,909	18,720	348,782	19,522	350,710	19,630	3.2	3.0	3	28	70
Fiji	1,895	2,440	2,068	2,638	4,861	6,200	2.1	3.2	20	21	59
New Zealand	51,655	14,340	57,070	16,026	61,214	17,190	1.4	1.9	8	26	66
Papua New Guinea	4,976	1,160	4,901	1,139	11,183	2,600	1.1	4.5	26	38	34
Solomon Islands	341	910	357	944	843	2,230	6.9	6.1	X	X	X

Sources: World Resources Institute, *World Resources, 1998–1999* and World Bank, *World Development Indicators, 1998.*

Table D

World Countries: Population Growth, 1950–2050

COUNTRY	POPULATION (thousands)				AVERAGE ANNUAL POPULATION CHANGE (percent)			AVERAGE ANNUAL INCREMENT TO THE POPULATION (thousands)		
	1950	1998	2025	2050	1985–90	1995–00	2005–10	1985–90	1995–00	2005–10
WORLD	2,513,878	5,929,839	8,039,130	9,366,724	1.7	1.4	1.2	86,966	80,848	80,011
AFRICA	223,974	778,484	1,453,899	2,046,401	2.8	2.6	2.5	16,330	20,063	24,232
Algeria	8,753	30,175	47,322	58,991	2.6	2.3	1.9	610	698	690
Angola	4,131	11,967	25,547	38,897	2.9	3.3	2.9	245	393	461
Benin	2,046	5,681	12,276	18,095	3.0	2.8	2.9	132	163	227
Botswana	389	1,551	2,576	3,320	3.3	2.2	2.0	39	34	39
Burkina Faso	3,654	11,402	23,451	35,419	2.8	2.8	2.8	240	316	414
Burundi	2,456	6,589	12,341	16,937	2.9	2.8	2.4	148	182	205
Cameroon	4,466	14,323	28,521	41,951	2.8	2.7	2.7	301	387	496
Central African Republic	1,314	3,489	6,006	8,215	2.4	2.1	2.1	66	73	89
Chad	2,658	6,892	12,648	18,004	2.0	2.8	2.3	106	187	200
Congo (Zaire)	12,184	49,208	105,925	164,635	3.3	2.6	3.0	1,142	1,259	1,943
Congo Republic	808	2,822	5,747	8,729	3.0	2.8	2.7	62	78	99
Cote d'Ivoire	2,776	14,567	24,397	31,706	3.3	2.0	2.2	349	290	395
Egypt	21,834	65,675	96,766	115,480	2.5	1.9	1.6	1,313	1,205	1,197
Equatorial Guinea	226	430	798	1,144	2.4	2.5	2.4	8	10	13
Eritrea	1,140	3,548	6,504	8,808	1.3	3.7	2.3	37	128	104
Ethiopia	18,434	62,111	136,288	212,732	3.1	3.2	3.0	1,401	1,954	2,480
Gabon	469	1,170	2,118	2,952	3.1	2.8	2.3	26	32	34
Gambia	294	1,194	1,984	2,604	4.2	2.3	2.0	34	27	29
Ghana	4,900	18,857	36,341	51,205	3.1	2.8	2.6	436	518	636
Guinea	2,550	7,673	15,286	22,914	2.9	1.4	2.8	154	102	272
Guinea-Bissau	505	1,134	1,921	2,674	2.0	2.0	2.0	18	22	28
Kenya	6,265	29,020	50,202	66,054	3.3	2.2	2.4	721	638	880
Lesotho	734	2,184	4,031	5,643	2.6	2.5	2.4	44	53	67
Liberia	824	2,748	6,573	9,955	3.2	8.6	3.0	75	227	123
Libya	1,029	5,980	12,885	19,109	3.7	3.3	3.0	152	196	246
Madagascar	4,229	16,348	34,476	50,807	3.4	3.1	3.0	395	504	643
Malawi	2,881	10,377	20,391	39,825	5.1	2.4	2.6	417	262	341
Mali	3,520	11,832	24,575	36,817	3.0	3.0	2.8	259	353	441
Mauritania	825	2,453	4,443	6,007	2.5	2.5	2.4	47	62	73
Mauritius	493	1,154	1,481	1,654	0.8	1.1	1.0	8	12	13
Morocco	8,953	28,012	39,925	47,276	2.1	1.8	1.4	479	492	437
Mozambique	6,198	18,691	35,444	51,774	0.9	2.5	2.5	128	461	582
Namibia	511	1,653	2,999	4,167	2.7	2.4	2.3	35	39	48
Niger	2,400	10,119	22,385	24,576	3.1	3.3	3.1	225	331	418
Nigeria	21,935	121,773	238,397	338,510	2.9	2.8	2.6	2,617	3,413	4,152
Rwanda	2,120	6,528	12,981	16,937	2.8	7.9	2.4	180	498	215
Senegal	2,500	9,001	16,896	23,442	2.8	2.7	2.5	190	237	286
Sierra Leone	1,944	4,577	8,200	11,368	2.2	3.0	2.2	82	134	125
Somalia	3,072	10,653	23,669	36,408	1.8	3.9	3.1	150	408	445
South Africa	13,683	44,295	71,621	91,466	2.3	2.2	2.0	805	958	1,050
Sudan	9,190	28,526	46,850	59,947	2.3	2.2	2.1	520	623	721
Swaziland	264	931	1,675	2,228	2.7	2.8	2.4	19	25	28
Tanzania	7,886	32,189	62,436	88,963	3.1	2.3	2.6	740	732	1,088
Togo	1,329	4,434	8,762	12,655	3.0	2.7	2.6	99	118	150
Tunisia	3,530	9,497	13,524	15,907	2.1	1.8	1.4	166	170	151
Uganda	4,762	21,318	44,983	66,305	2.4	2.6	2.9	377	554	821
Zambia	2,440	8,690	16,163	21,965	2.4	2.5	2.5	163	210	275
Zimbabwe	2,730	11,924	19,347	24,904	3.2	2.1	2.0	293	237	293
NORTH AND CENTRAL AMERICA										
Belize	69	230	375	480	2.4	2.5	2.0	4	6	6
Canada	13,737	30,194	36,385	36,352	1.4	0.9	0.7	370	255	231
Costa Rica	862	3,650	5,608	6,902	2.8	2.1	1.7	79	75	74
Cuba	5,850	11,115	11,798	13,141	1.0	0.4	0.3	103	47	29
Dominican Republic	2,353	8,232	11,164	13,141	2.2	1.7	1.2	147	134	117

(Continued on next page)

COUNTRY	POPULATION (thousands)				AVERAGE ANNUAL POPULATION CHANGE (percent)			AVERAGE ANNUAL INCREMENT TO THE POPULATION (thousands)		
	1950	1998	2025	2050	1985–90	1995–00	2005–10	1985–90	1995–00	2005–10
El Salvador	1,951	6,059	9,221	11,364	1.5	2.2	1.7	71	131	121
Guatemala	2,969	11,562	21,668	29,353	2.9	2.8	2.5	247	320	371
Haiti	3,261	7,534	12,513	17,524	2.0	1.9	1.9	122	139	169
Honduras	1,380	6,147	10,656	13,920	3.1	2.8	2.2	139	166	171
Jamaica	1,403	2,539	3,370	3,886	0.5	0.9	1.2	11	24	32
Mexico	27,727	95,831	130,196	154,120	2.0	1.6	1.2	1,552	1,547	1,349
Nicaragua	1,098	4,464	7,639	9,922	2.2	2.6	2.2	73	114	123
Panama	860	2,767	3,779	4,365	2.0	1.6	1.3	46	45	40
Trinidad and Tobago	636	1,318	1,692	1,899	1.0	0.8	1.1	12	11	15
United States	157,813	273,754	332,481	347,543	1.0	0.8	0.8	2,450	2,142	2,204

SOUTH AMERICA

COUNTRY	1950	1998	2025	2050	1985–90	1995–00	2005–10	1985–90	1995–00	2005–10
Argentina	17,150	36,123	47,160	54,522	1.4	1.3	1.1	444	453	434
Bolivia	2,714	7,957	13,131	16,956	2.2	2.3	2.0	136	183	191
Brazil	53,975	165,158	216,596	243,259	1.8	1.2	1.1	2,548	2,037	2,067
Chile	6,082	14,824	19,548	22,215	1.7	1.4	1.1	210	200	175
Colombia	11,945	37,685	52,668	62,284	2.1	1.7	1.3	636	618	579
Ecuador	3,387	12,175	17,796	21,290	2.4	2.0	1.5	233	237	220
Guyana	423	856	1,114	1,239	0.1	1.0	1.0	0	9	10
Paraguay	1,488	5,222	9,355	12,565	3.1	2.6	2.3	122	134	153
Peru	7,632	24,797	35,518	42,292	2.0	1.7	1.4	415	426	416
Suriname	215	442	605	711	1.2	1.2	1.3	5	5	6
Uruguay	2,239	3,239	3,692	4,027	0.6	0.6	0.5	17	18	18
Venezuela	5,094	23,242	34,775	42,152	2.6	2.0	1.6	473	465	450

ASIA

COUNTRY	1950	1998	2025	2050	1985–90	1995–00	2005–10	1985–90	1995–00	2005–10
Afghanistan	8,958	23,364	45,262	61,373	0.3	5.3	2.4	47	1,186	737
Armenia	1,354	3,646	4,185	4,376	1.2	0.2	0.7	41	6	25
Azerbaijan	2,896	7,714	9,714	10,881	1.4	0.8	0.9	98	59	72
Bangladesh	41,783	124,043	179,980	218,188	2.0	1.6	1.6	2,091	2,016	2,396
Bhutan	734	1,917	3,646	5,184	2.5	2.8	2.5	39	52	60
Cambodia	4,346	10,751	16,990	21,394	3.2	2.2	1.8	255	237	226
China	554,760	1,255,091	1,480,430	1,516,664	1.5	0.9	0.7	17,026	11,215	8,676
Georgia	3,527	5,428	5,762	6,028	0.6	(0.1)	0.2	35	(6)	11
India	357,561	975,772	1,330,201	1,532,674	2.1	1.6	1.3	16,571	15,553	14,020
Indonesia	79,538	206,522	275,245	318,264	1.8	1.5	1.1	3,096	3,021	2,488
Iran	16,913	73,057	128,251	170,269	3.8	2.2	2.4	2,061	1,613	2,250
Iraq	5,158	21,795	41,600	56,129	3.3	2.8	2.6	552	603	751
Israel	1,258	5,883	7,977	9,144	1.9	1.9	1.2	85	110	83
Japan	83,625	125,920	121,348	109,546	0.4	0.2	(0.0)	540	272	(30)
Jordan	1,237	5,956	11,894	16,671	2.1	3.3	2.8	85	191	217
Kazakhstan	6,703	16,854	20,047	22,260	1.1	0.1	0.7	183	22	121
Korea, North	9,488	23,206	30,046	32,873	1.5	1.6	0.9	284	363	232
Korea, South	20,357	46,115	52,533	52,146	1.0	0.9	0.6	413	395	288
Kuwait	152	1,809	2,904	3,406	4.4	3.0	1.7	85	55	40
Kyrgyzstan	1,740	4,497	5,950	7,182	1.8	0.4	1.1	76	17	51
Laos	1,755	5,358	10,202	13,889	3.1	3.1	2.6	122	162	181
Lebanon	1,443	3,194	4,424	5,189	(0.9)	1.8	1.1	(23)	56	41
Malaysia	6,110	21,450	32,577	38,089	2.6	2.0	1.5	443	432	382
Mongolia	761	2,624	4,052	4,986	3.0	2.1	1.8	61	55	58
Myanmar (Burma)	17,832	47,625	67,654	80,896	1.9	1.8	1.4	762	847	796
Nepal	7,862	23,168	40,554	53,621	2.6	2.5	2.3	454	578	657
Oman	456	2,504	6,538	10,930	4.5	4.2	3.8	72	102	137
Pakistan	39,513	147,811	268,904	357,353	3.3	2.7	2.4	3,589	3,950	4,606
Philippines	20,988	72,164	105,194	130,511	2.1	2.0	1.6	122	1,440	1,342
Saudi Arabia	3,201	20,207	43,363	59,812	4.8	2.4	2.9	680	681	793
Singapore	1,022	3,491	4,212	4,190	2.2	1.5	0.7	61	52	27
Sri Lanka	7,678	18,450	23,934	26,995	1.2	1.0	1.1	199	179	224
Syria	3,495	15,335	26,303	34,463	3.5	2.5	2.3	398	385	446
Tajikistan	1,532	6,161	9,747	12,366	3.0	1.9	1.9	147	114	141
Thailand	20,010	59,612	69,089	72,969	1.7	0.8	0.6	890	451	391
Turkey	20,809	63,763	85,791	97,911	2.2	1.6	1.2	1,151	979	834
Turkmenistan	1,211	4,316	6,470	7,916	2.6	1.9	1.7	88	81	85
United Arab Emirates	70	2,354	3,297	3,668	4.3	2.0	1.5	74	47	42

(Continued on next page)

	POPULATION (thousands)				AVERAGE ANNUAL POPULATION CHANGE (percent)			AVERAGE ANNUAL INCREMENT TO THE POPULATION (thousands)		
COUNTRY	1950	1998	2025	2050	1985–90	1995–00	2005–10	1985–90	1995–00	2005–10
Uzbekistan	6,314	24,105	36,500	45,094	2.4	1.9	1.7	468	451	495
Vietnam	29,954	77,896	110,107	129,763	2.2	1.8	1.2	1,358	1,351	1,083
Yemen	4,316	16,891	39,589	61,129	3.6	3.7	3.3	379	618	775
EUROPE										
Albania	1,230	3,445	4,295	4,747	2.1	0.6	0.8	65	22	28
Austria	6,935	8,210	8,305	7,430	0.4	0.6	0.1	29	49	9
Belarus	7,745	10,323	9,641	8,726	0.5	(0.1)	(0.2)	52	(14)	(22)
Belgium	8,639	10,213	10,271	9,763	0.2	0.3	0.0	19	26	4
Bosnia-Herzegovina	2,661	3,994	4,303	3,789	0.9	3.9	0.0	37	154	1
Bulgaria	7,251	8,387	7,453	6,690	(0.6)	(0.5)	(0.4)	(48)	(41)	(31)
Croatia	3,850	4,494	4,243	3,991	0.2	(0.1)	(0.2)	9	(4)	(8)
Czech Republic	8,925	10,223	9,627	8,572	0.0	(0.1)	(0.2)	0	(14)	(16)
Denmark	4,271	5,258	5,324	5,234	0.1	0.2	0.0	5	10	1
Estonia	1,101	1,442	1,256	1,084	0.7	(1.0)	(0.4)	10	(14)	(6)
Finland	4,009	5,156	5,294	5,172	0.3	0.3	0.1	17	14	5
France	41,829	58,733	60,393	58,370	0.6	0.3	0.1	310	191	67
Germany	68,376	82,401	80,877	69,542	0.4	0.3	(0.1)	339	219	(57)
Greece	7,566	10,551	10,074	9,013	0.6	0.3	(0.2)	57	29	(17)
Hungary	9,338	9,930	8,667	7,715	(0.4)	(0.6)	(0.5)	(43)	(59)	(48)
Iceland	143	277	336	363	1.1	1.0	0.8	3	3	2
Ireland	2,969	3,564	3,723	3,809	(0.3)	0.2	0.3	(10)	6	11
Italy	47,104	57,244	51,774	42,092	0.1	0.0	(0.3)	50	(2)	(175)
Latvia	1,949	2,447	2,108	1,891	0.7	(1.1)	(0.5)	18	(28)	(12)
Lithuania	2,567	3,710	3,521	3,297	1.0	(0.3)	(0.2)	36	(9)	(6)
Macedonia	1,230	2,205	2,541	2,646	1.2	0.7	0.6	25	15	14
Moldova	2,341	4,451	4,869	5,138	0.7	0.1	0.4	30	4	18
Netherlands	10,114	15,739	16,141	14,956	0.6	0.5	0.1	92	78	20
Norway	3,265	4,379	4,662	4,696	0.4	0.4	0.2	18	15	10
Poland	24,824	38,664	39,973	39,725	0.5	0.1	0.2	183	34	77
Portugal	8,405	9,798	9,438	8,701	(0.1)	(0.1)	(0.1)	(7)	(5)	(8)
Romania	16,311	22,573	21,098	19,009	0.4	(0.2)	(0.2)	96	(45)	(47)
Russian Federation	102,192	147,231	131,995	114,318	0.7	(0.3)	(0.4)	993	(453)	(513)
Slovak Republic	3,463	5,360	5,469	5,260	0.5	0.1	0.2	23	7	8
Slovenia	1,473	1,919	1,738	1,471	0.4	(0.1)	(0.3)	7	(2)	(5)
Spain	28,009	39,754	37,500	31,755	0.4	0.1	(0.1)	160	35	(57)
Sweden	7,014	8,863	9,511	9,574	0.5	0.3	0.2	42	22	22
Switzerland	4,694	7,325	7,581	6,935	0.9	0.7	0.2	60	49	15
Ukraine	37,906	51,218	45,979	40,802	0.4	(0.4)	(0.4)	190	(191)	(175)
United Kingdom	50,616	58,249	59,535	58,733	0.3	0.1	0.1	189	51	37
Yugoslavia	7,131	10,410	10,679	10,979	0.6	0.5	(0.0)	62	50	(1)
OCEANIA										
Australia	8,219	18,445	29,931	25,286	1.5	1.1	1.0	249	194	201
Fiji	289	822	1,170	1,393	0.8	1.6	1.5	5	13	14
New Zealand	1,908	3,680	4,878	5,271	0.7	1.1	1.1	23	40	43
Papua New Guinea	1,613	4,602	7,546	9,637	2.2	2.2	2.0	79	102	113
Solomon Islands	90	417	844	1,192	3.4	3.2	2.8	10	13	16
DEVELOPING COUNTRIES	1,711,191	4,783,310	6,818,880	8,204,983	2.0	1.7	1.4	80,235	77,726	78,204
DEVELOPED COUNTRIES	812,687	1,181,530	1,220,250	1,161,741	0.5	0.3	0.2	6,761	3,121	1,806

Source: United Nations Population Division and International Labour Organisation. World Resources Institute, *World Resources 1998–99.*

Table E

World Countries: Basic Demographic Data, 1975–2000

COUNTRY	CRUDE BIRTH RATE (births per 1,000 population)		LIFE EXPECTANCY AT BIRTH (years)		LIFE EXPECTANCY OF FEMALES AS A PERCENTAGE OF MALES (years)		TOTAL FERTILITY RATE		PERCENTAGE OF POPULATION IN SPECIFIC AGE GROUPS					
									1980			2000		
	1975–1980	1995–00	1975–80	1995–00	1975–80	1995–00	1975–80	1995–00	<15	15–65	>65	<15	15–65	>65
WORLD	28.3	22.6	59.7	65.6	106.0	106.8	3.9	2.8	35.2	58.9	5.9	30.0	63.2	6.8
AFRICA	**46.0**	**39.2**	**47.9**	**53.8**	**106.7**	**105.7**	**6.5**	**5.3**	**44.7**	**52.2**	**3.1**	**43.0**	**53.8**	**3.2**
Algeria	45.0	29.2	57.5	68.9	103.5	104.1	7.2	3.8	46.5	49.6	3.9	36.6	59.6	3.8
Angola	50.2	47.7	40.0	46.5	108.1	107.1	6.8	6.7	44.7	52.4	2.9	47.4	49.8	2.8
Benin	51.4	42.0	47.0	54.8	108.2	109.2	7.1	5.8	45.1	50.8	4.1	46.5	50.7	2.8
Botswana	46.6	235.0	56.5	50.4	106.6	105.7	6.4	4.5	48.7	49.3	2.0	41.9	55.6	2.4
Burkina Faso	50.8	45.9	42.9	46.0	106.0	104.2	7.8	6.6	47.4	49.8	2.8	47.0	50.3	2.7
Burundi	44.7	42.5	46.0	47.2	107.2	107.3	6.8	6.3	44.7	51.8	3.4	45.0	52.3	2.7
Cameroon	45.5	39.3	48.5	55.9	106.4	105.0	6.5	5.3	44.4	52.0	3.6	43.5	52.9	3.6
Central African Republic	44.1	37.6	44.5	48.6	111.9	109.9	5.9	5.0	41.7	54.4	4.0	41.6	54.4	4.0
Chad	44.1	41.6	41.0	47.7	108.1	106.5	5.9	5.5	41.9	54.5	3.6	43.0	53.4	3.6
Congo (Zaire)	47.8	44.9	48.0	52.9	107.1	106.2	6.5	6.2	46.0	51.1	2.8	48.0	49.1	2.9
Congo Republic	45.8	42.5	48.7	50.9	111.3	109.9	6.3	5.9	45.1	51.5	3.4	45.7	51.1	3.3
Cote d'Ivoire	50.7	37.2	47.9	51.0	107.1	104.4	7.4	5.1	46.6	51.0	2.5	43.0	54.0	3.0
Egypt	38.9	26.1	54.1	66.0	104.5	104.0	5.3	3.4	39.5	56.5	4.0	35.0	60.5	4.5
Equatorial Guinea	42.7	40.8	42.0	50.0	107.9	106.6	5.7	5.5	41.0	54.8	4.1	43.1	52.9	4.0
Eritrea	45.1	39.8	45.3	50.6	106.8	106.1	6.1	5.3	44.2	53.1	2.6	43.6	53.3	3.1
Ethiopia	49.0	48.2	42.0	49.9	107.9	106.6	6.8	7.0	46.1	51.2	2.7	47.1	50.2	2.8
Gabon	32.9	37.6	47.0	55.5	107.3	106.3	4.4	5.4	34.4	59.5	6.1	39.8	54.6	5.7
Gambia	48.8	39.9	39.0	47.0	108.3	107.3	6.5	5.2	42.6	54.4	3.1	41.2	55.5	3.1
Ghana	45.1	38.2	51.0	58.0	106.9	106.6	6.5	5.3	44.9	52.3	2.8	43.4	53.6	3.0
Guinea	51.6	48.2	38.8	46.5	102.6	102.2	7.0	6.6	45.8	51.6	2.6	47.0	50.4	2.6
Guinea-Bissau	42.4	40.3	37.5	43.8	108.6	106.6	5.6	5.4	39.0	57.0	4.0	41.7	54.2	4.2
Kenya	53.6	36.9	53.4	54.5	107.8	106.5	8.1	4.9	50.1	46.5	3.4	43.4	53.7	2.9
Lesotho	41.9	35.4	51.8	58.6	107.0	104.5	5.7	4.9	41.9	53.9	4.2	41.0	55.0	4.1
Liberia	47.4	47.5	49.5	51.5	106.3	106.0	6.8	6.3	44.3	52.0	3.7	43.7	52.7	3.6
Libya	47.3	40.0	55.8	65.5	106.3	105.6	7.4	5.9	46.7	51.1	2.2	44.7	52.4	2.9
Madagascar	46.9	41.1	49.5	58.5	106.3	105.3	6.6	5.7	45.9	51.4	2.7	45.8	51.6	2.6
Malawi	57.2	47.7	43.1	40.7	103.3	102.0	7.6	6.7	47.5	50.3	2.3	46.4	50.9	2.7
Mali	50.7	47.4	45.0	48.0	108.1	107.1	7.1	6.6	46.8	50.7	2.5	47.2	50.2	2.5
Mauritania	44.7	38.3	45.5	53.5	107.3	106.2	6.5	5.0	43.7	53.3	3.0	41.5	55.2	3.3
Mauritius	26.7	19.3	64.9	71.6	108.3	109.8	3.1	2.3	35.6	60.8	3.6	26.6	67.3	6.0
Morocco	39.4	25.8	55.8	66.7	106.3	105.7	5.9	3.1	43.2	52.7	4.1	33.8	61.9	4.3
Mozambique	45.4	42.5	43.5	46.9	107.6	106.4	6.5	6.1	43.4	53.4	3.2	44.7	52.1	3.2
Namibia	41.9	35.9	51.3	55.5	105.0	103.5	6.0	4.9	43.1	53.4	3.5	41.6	54.6	3.8
Niger	59.7	50.2	40.5	48.5	108.2	107.0	8.1	7.1	46.8	50.8	2.5	48.6	49.0	2.4
Nigeria	46.6	42.3	45.0	52.4	107.4	106.3	6.5	6.0	44.3	53.1	2.6	45.0	52.1	2.9
Rwanda	52.8	42.8	45.0	42.1	107.4	106.4	8.5	6.0	48.8	48.8	2.4	44.7	53.0	2.4
Senegal	49.3	41.1	42.8	51.3	104.8	104.0	7.0	5.6	45.3	51.8	2.8	43.6	53.4	3.0
Sierra Leone	48.8	46.5	35.2	37.5	108.9	108.6	6.5	6.1	43.0	53.9	3.1	43.9	53.1	3.0
Somalia	50.4	50.0	42.0	49.0	107.9	106.8	7.0	7.0	46.0	51.0	3.0	48.0	49.5	2.6
South Africa	37.3	29.7	55.9	65.2	111.3	109.6	5.1	3.8	40.3	55.9	3.9	36.2	59.3	4.5
Sudan	47.1	33.7	46.7	55.0	106.2	105.2	6.7	4.6	44.9	52.4	2.7	38.7	58.1	3.2
Swaziland	46.2	36.8	49.9	60.0	109.9	108.0	6.5	4.5	45.9	51.3	2.9	41.7	55.6	2.7

(Continued on next page)

COUNTRY	CRUDE BIRTH RATE (births per 1,000 population)		LIFE EXPECTANCY AT BIRTH (years)		LIFE EXPECTANCY OF FEMALES AS A PERCENTAGE OF MALES (years)		TOTAL FERTILITY RATE		PERCENTAGE OF POPULATION IN SPECIFIC AGE GROUPS					
									1980			2000		
	1975–1980	1995–00	1975–80	1995–00	1975–80	1995–00	1975–80	1995–00	<15	15–65	>65	<15	15–65	>65
Tanzania	47.5	41.2	49.0	51.4	107.2	105.6	6.8	5.5	47.6	50.1	2.3	45.1	52.3	2.6
Togo	45.2	41.9	48.0	50.1	107.1	105.5	6.6	6.1	44.6	52.3	3.2	45.6	51.3	3.1
Tunisia	36.4	23.9	60.0	69.6	101.7	103.4	5.7	2.9	41.6	54.6	3.8	32.2	62.9	4.9
Uganda	50.3	51.1	47.0	41.4	107.0	104.7	6.9	7.1	47.8	49.7	2.5	49.1	48.6	2.2
Zambia	51.6	42.4	49.3	43.0	106.9	103.6	7.2	5.5	49.4	48.2	2.4	46.3	51.4	2.3
Zimbabwe	44.2	37.1	53.8	48.5	106.9	103.8	6.6	4.7	47.9	49.5	2.6	43.6	53.7	2.7
NORTH AMERICA	**15.1**	**13.6**	**73.3**	**76.9**	**111.2**	**109.1**	**1.8**	**1.9**	**22.5**	**66.4**	**11.0**	**21.2**	**66.4**	**12.4**
Canada	15.4	11.9	74.2	78.9	110.8	107.5	1.8	1.6	22.7	67.9	9.4	19.3	68.1	12.6
United States	15.1	13.8	73.2	76.7	111.2	109.1	1.8	2.0	22.5	66.3	11.2	21.4	66.2	12.4
CENTRAL AMERICA	**38.2**	**26.5**	**63.7**	**71.7**	**110.2**	**108.4**	**5.4**	**3.0**	**45.1**	**51.2**	**3.6**	**34.8**	**60.7**	**4.5**
Belize	40.9	31.3	69.7	74.7	102.5	103.7	6.2	3.7	47.2	48.6	4.8	39.7	55.8	4.1
Costa Rica	31.7	24.0	71.0	76.8	106.4	106.3	3.9	3.0	38.9	57.5	3.6	33.1	61.8	5.1
Cuba	17.2	13.1	73.0	76.0	104.8	105.1	2.1	1.6	32.9	60.5	7.6	21.2	69.2	9.6
Dominican Republic	34.9	24.1	62.0	70.9	106.1	106.1	4.7	2.8	42.3	54.6	3.1	33.1	62.5	4.5
El Salvador	41.5	27.9	57.2	69.5	119.5	109.0	5.7	3.1	46.1	50.8	3.1	35.6	59.7	4.7
Guatemala	44.3	36.3	56.4	67.2	107.2	107.9	6.4	4.9	45.9	51.3	2.8	42.9	53.3	3.7
Haiti	36.8	34.1	50.7	54.5	106.3	106.1	5.4	4.6	40.7	54.8	4.4	40.0	56.2	3.8
Honduras	44.9	33.5	57.7	69.8	107.7	107.1	6.6	4.3	47.2	50.1	2.7	41.6	54.9	3.4
Jamaica	28.8	21.7	70.1	74.6	106.3	106.1	4.0	2.4	40.2	53.0	6.8	30.2	53.4	6.4
Mexico	37.1	24.6	65.3	72.5	110.3	108.6	5.3	2.8	45.1	51.1	3.8	33.1	62.1	4.7
Nicaragua	45.7	33.5	57.6	68.2	108.5	107.3	6.4	3.9	47.7	49.8	2.5	40.8	56.0	3.2
Panama	31.0	22.5	69.0	73.9	105.8	106.4	4.1	2.6	40.5	55.0	4.5	31.3	63.7	5.5
Trinidad and Tobago	29.3	16.6	68.3	73.7	107.6	106.6	3.4	2.1	34.2	60.2	5.5	26.1	67.4	6.5
SOUTH AMERICA	**32.0**	**21.5**	**62.9**	**69.0**	**108.8**	**110.7**	**4.3**	**2.5**	**37.8**	**57.6**	**4.6**	**30.2**	**64.2**	**5.6**
Argentina	25.7	19.9	68.7	73.2	110.4	110.3	3.4	2.6	30.5	61.4	8.1	27.7	62.6	9.7
Bolivia	41.0	33.2	50.1	61.5	108.8	105.7	5.8	4.4	42.6	53.9	3.5	30.5	56.4	4.0
Brazil	32.6	10.6	61.8	67.1	108.1	112.3	4.3	2.2	38.1	57.8	4.2	28.4	66.4	5.2
Chile	24.0	19.9	67.2	75.3	110.5	108.3	3.0	2.4	33.5	60.9	5.6	28.5	64.4	7.2
Colombia	31.7	23.4	64.0	70.9	107.3	108.1	4.1	2.7	40.0	56.2	3.7	32.5	62.8	4.6
Ecuador	38.2	25.6	61.4	69.8	105.9	107.7	5.4	3.1	42.8	53.2	4.0	33.8	61.5	4.7
Guyana	31.5	21.9	60.7	64.4	108.4	111.1	3.9	2.3	40.8	55.2	4.0	29.9	65.9	4.2
Paraguay	35.9	32.3	66.6	69.7	106.7	106.7	5.2	4.2	42.2	53.3	4.5	39.5	57.0	3.5
Peru	38.0	24.9	58.5	68.3	106.7	107.6	5.4	3.0	41.9	54.5	3.6	33.4	61.8	4.8
Suriname	29.5	21.8	65.1	71.5	107.8	107.2	4.2	2.4	39.7	55.8	4.5	32.3	62.4	5.5
Uruguay	20.3	16.8	69.6	72.9	110.2	109.3	2.9	2.3	26.9	62.5	10.5	23.9	63.5	12.7
Venezuela	34.2	24.9	67.6	72.8	109.1	108.1	4.5	3.0	40.7	56.1	3.2	34.0	61.5	4.4
ASIA	**29.6**	**22.3**	**58.5**	**66.2**	**102.6**	**104.5**	**4.2**	**2.7**	**37.6**	**58.0**	**4.4**	**30.1**	**64.1**	**5.8**
Afghanistan	50.8	53.4	40.0	45.5	100.	102.2	7.2	6.9	43.0	54.5	2.5	41.7	55.6	2.7
Armenia	20.9	13.3	72.3	70.6	109.0	110.1	2.5	1.7	30.4	63.6	6.0	24.5	66.8	8.7
Azerbaijan	24.7	19.2	68.5	70.6	112.1	112.0	3.6	2.3	34.5	60.0	5.4	29.5	63.6	6.9
Bangladesh	47.2	26.8	46.6	58.1	97.9	100.2	6.7	3.1	46.0	50.5	3.4	35.6	61.2	3.3
Bhutan	42.8	41.3	42.6	53.2	104.8	106.4	5.9	5.9	41.3	55.6	3.1	43.0	53.8	3.2
Cambodia	30.0	33.7	31.2	54.1	108.3	105.3	4.1	4.5	39.2	57.9	2.9	40.4	56.6	3.0
China	21.5	16.2	65.3	69.9	103.0	105.1	3.3	1.8	35.5	59.7	4.7	24.9	68.4	6.7

(Continued on next page)

COUNTRY	CRUDE BIRTH RATE (births per 1,000 population)		LIFE EXPECTANCY AT BIRTH (years)		LIFE EXPECTANCY OF FEMALES AS A PERCENTAGE OF MALES (years)		TOTAL FERTILITY RATE		PERCENTAGE OF POPULATION IN SPECIFIC AGE GROUPS					
									1980			2000		
	1975–1980	1995–00	1975–80	1995–00	1975–80	1995–00	1975–80	1995–00	<15	15–65	>65	<15	15–65	>65
Georgia	18.1	13.8	70.7	72.7	111.9	112.0	2.4	1.9	25.7	65.1	9.1	22.0	65.4	12.6
India	34.7	25.2	52.9	62.4	96.3	101.0	4.8	3.1	38.5	57.4	4.0	32.7	62.3	5.0
Indonesia	35.4	23.1	52.8	65.1	104.9	105.8	4.7	2.6	41.0	55.6	3.3	30.8	64.6	4.7
Iran	44.7	34.0	58.6	69.2	101.4	102.2	6.5	4.8	44.9	51.8	3.3	43.1	52.8	4.1
Iraq	41.9	36.4	61.4	62.4	103.0	104.9	6.6	5.3	46.0	51.3	2.7	41.4	55.5	3.1
Israel	26.0	20.3	73.1	77.7	104.9	105.0	3.4	2.8	33.2	58.2	8.6	28.1	62.4	9.5
Japan	15.2	10.3	75.5	80.0	107.4	107.8	1.8	1.5	23.6	67.4	9.0	15.2	68.3	16.5
Jordan	45.0	37.5	61.2	69.7	106.1	106.1	7.4	5.1	49.4	47.5	3.1	43.3	53.8	2.9
Kazakhstan	24.9	18.1	65.4	67.6	117.1	115.4	3.1	2.3	32.4	61.5	6.1	27.5	65.4	7.1
Korea, North	22.2	21.3	65.8	72.2	110.3	109.0	3.3	2.1	39.5	57.0	3.5	27.3	57.4	5.3
Korea, South	23.9	15.0	64.8	72.4	111.6	110.5	2.9	1.7	34.0	62.2	3.8	21.4	72.0	6.7
Kuwait	40.1	21.3	69.6	76.0	106.2	105.5	5.9	2.8	40.2	58.3	1.4	33.2	64.8	2.0
Kyrgyzstan	29.9	25.5	64.2	67.6	114.2	113.4	4.1	3.2	37.1	57.1	5.8	35.0	59.0	6.0
Laos	45.1	44.2	43.5	53.5	106.9	105.8	6.7	6.7	42.0	55.2	2.8	45.4	51.7	3.0
Lebanon	30.1	24.2	65.0	69.9	106.2	105.3	4.3	2.8	40.1	54.5	5.4	32.9	61.3	5.8
Malaysia	30.4	25.2	65.3	72.0	105.7	106.3	4.2	3.2	39.3	57.0	3.7	35.3	60.6	4.1
Mongolia	39.2	27.8	56.3	65.8	104.5	104.7	6.7	3.3	43.2	53.9	2.9	36.4	59.8	3.8
Myanmar (Burma)	37.5	27.2	51.2	60.1	106.4	105.6	5.3	3.3	39.6	56.4	4.0	34.0	61.4	4.6
Nepal	43.4	36.3	46.2	57.3	96.6	99.1	6.2	5.0	42.9	54.1	3.0	42.0	54.5	3.5
Oman	46.1	44.1	54.9	70.8	104.3	106.4	7.2	7.2	44.6	52.8	2.6	47.7	49.9	2.3
Pakistan	47.3	36.1	53.4	63.9	101.3	103.5	7.0	5.0	44.4	52.7	2.9	41.8	55.0	3.2
Philippines	35.9	28.4	59.9	68.3	105.5	105.4	5.0	3.6	41.9	55.3	2.8	36.7	59.7	3.6
Saudi Arabia	45.9	34.3	58.8	71.4	104.0	105.0	7.3	5.9	44.3	52.9	2.8	40.7	56.4	2.9
Singapore	17.2	15.7	70.8	77.3	106.6	105.9	1.9	1.8	27.0	68.2	4.7	22.6	70.3	7.1
Sri Lanka	28.5	17.8	66.8	73.1	105.4	106.3	3.8	2.1	35.3	60.4	4.3	26.2	67.2	6.6
Syria	46.0	30.4	60.1	68.9	106.2	106.7	7.4	4.0	48.5	48.3	3.2	40.8	56.1	3.1
Tajikistan	37.2	30.4	64.5	67.2	108.3	109.3	5.9	3.9	42.9	52.5	4.6	36.5	55.9	4.6
Thailand	31.6	16.7	61.2	69.3	106.6	109.0	4.3	1.7	40.0	56.5	3.5	25.2	69.1	5.8
Turkey	32.0	21.9	60.3	69.0	107.8	107.8	4.5	2.5	39.2	56.0	4.7	28.3	65.7	5.9
Turkmenistan	35.3	28.6	61.6	64.6	11.20	111.1	5.3	3.6	41.3	54.4	4.3	37.4	58.3	4.2
United Arab Emirates	30.5	18.7	66.8	74.8	106.5	103.5	5.7	3.5	28.6	70.1	1.3	28.1	69.4	2.5
Uzbekistan	34.4	28.2	65.1	67.5	111.0	110.0	5.1	3.5	40.9	54.0	5.1	37.5	57.9	4.6
Viet Nam	38.3	25.1	55.8	67.4	108.2	107.2	5.6	3.0	42.5	52.7	4.8	34.3	60.5	5.2
Yemen	53.7	47.7	44.1	58.0	101.1	101.7	7.6	7.6	50.2	47.2	2.6	48.3	49.3	2.3
EUROPE	**14.8**	**10.5**	**71.3**	**72.6**	**111.4**	**112.7**	**2.0**	**1.5**	**22.2**	**65.5**	**12.3**	**17.5**	**67.9**	**14.6**
Albania	30.3	21.4	68.9	70.9	106.6	106.8	4.2	2.6	35.9	58.9	5.2	29.7	64.4	6.0
Austria	11.5	10.3	72.0	77.0	110.4	108.7	1.7	1.4	20.4	64.2	15.4	17.1	68.5	14.4
Belarus	16.1	10.0	71.1	69.6	114.7	116.1	2.1	1.4	22.9	66.4	10.7	18.8	67.3	13.8
Belgium	12.4	11.2	72.3	77.3	109.6	109.1	1.7	1.6	20.2	65.5	14.3	17.4	66.2	16.4
Bosnia-Herzegovina	19.6	10.8	69.9	73.2	107.0	107.7	2.2	1.4	27.3	66.7	6.1	18.9	71.2	9.8
Bulgaria	15.8	10.3	71.3	71.3	107.9	110.5	2.2	1.5	22.1	66.0	11.9	16.9	67.2	15.8
Croatia	15.6	10.8	70.6	72.2	110.4	112.3	2.0	1.6	21.2	67.2	11.7	17.3	68.0	14.6
Czech Republic	17.4	10.7	70.6	72.9	110.4	108.9	2.3	1.4	23.5	63.2	13.4	17.6	69.8	12.6
Denmark	12.3	13.0	74.2	75.6	108.4	107.3	1.7	1.8	20.8	64.7	14.4	18.6	66.7	14.7
Estonia	15.0	9.0	69.7	69.4	115.3	117.4	2.1	1.3	21.7	65.8	12.5	17.6	68.5	13.9
Finland	13.6	12.0	72.7	76.6	112.6	109.7	1.6	1.8	20.3	67.7	12.0	18.4	67.1	14.6
France	14.0	11.6	73.7	76.8	111.6	111.1	1.9	1.6	22.3	63.8	14.0	18.3	65.4	16.2
Germany	10.4	9.3	72.5	76.7	109.4	108.9	1.5	1.3	18.5	65.9	15.6	15.4	68.8	15.9

(Continued on next page)

COUNTRY	CRUDE BIRTH RATE (births per 1,000 population)		LIFE EXPECTANCY AT BIRTH (years)		LIFE EXPECTANCY OF FEMALES AS A PERCENTAGE OF MALES (years)		TOTAL FERTILITY RATE		PERCENTAGE OF POPULATION IN SPECIFIC AGE GROUPS					
									1980			2000		
	1975–1980	1995–00	1975–80	1995–00	1975–80	1995–00	1975–80	1995–00	<15	15–65	>65	<15	15–65	>65
Greece	15.6	10.0	73.7	78.1	105.7	106.8	2.3	1.4	22.8	64.0	13.1	15.3	66.9	17.8
Hungary	16.2	10.2	69.4	69.0	109.8	114.4	2.1	1.4	21.9	64.6	13.4	17.0	68.5	14.5
Iceland	19.2	16.5	76.3	79.3	108.0	104.9	2.3	2.2	27.6	62.7	10.1	23.8	64.9	11.3
Ireland	21.2	13.0	72.0	76.6	107.2	107.3	3.5	1.8	30.6	58.7	10.7	21.2	67.5	11.4
Italy	13.0	9.1	73.6	78.3	109.2	108.4	1.9	1.2	22.3	64.6	13.1	14.2	68.1	17.7
Latvia	14.0	9.8	69.2	68.4	115.6	118.9	2.0	1.4	20.4	66.5	13.0	18.2	67.5	14.4
Lithuania	15.4	10.8	70.8	70.4	114.2	117.1	2.1	1.5	23.6	65.1	11.3	19.5	67.0	13.5
Macedonia	22.6	14.5	69.6	72.5	105.2	106.3	2.7	1.9	28.6	64.5	6.9	22.3	68.3	9.4
Moldova	19.6	13.6	64.8	67.5	111.2	112.6	2.4	1.8	26.7	65.5	7.8	23.4	66.8	9.8
Netherlands	12.7	11.9	75.3	77.9	109.0	107.5	1.6	1.6	22.3	66.2	11.5	18.2	68.2	13.6
Norway	12.9	13.4	75.3	77.6	108.9	107.8	1.8	1.9	22.2	63.1	14.8	19.8	65.1	15.0
Poland	19.2	11.9	70.9	71.1	111.9	113.5	2.3	1.7	24.3	65.6	10.1	19.8	68.4	11.8
Portugal	18.2	11.2	70.2	75.4	110.6	109.9	2.4	1.5	25.9	63.6	10.5	16.7	67.5	15.7
Romania	19.1	11.0	69.6	69.5	107.0	110.9	2.6	1.4	26.7	63.1	10.3	18.5	68.4	13.2
Russian Federation	15.8	9.6	67.4	64.4	118.1	123.3	1.9	1.4	21.6	68.1	10.2	18.0	69.3	12.7
Slovak Republic	20.6	11.7	70.4	71.3	110.9	113.1	2.5	1.5	26.1	63.5	10.4	20.0	68.8	11.1
Slovenia. Rep	16.6	9.5	71.0	73.5	111.6	112.4	2.2	1.3	23.4	65.3	11.4	15.7	70.2	14.2
Spain	17.4	9.7	74.3	78.0	108.4	109.4	2.6	1.2	26.6	62.7	10.7	15.0	68.4	16.5
Sweden	11.7	11.9	75.2	78.5	109.3	106.0	1.7	1.8	19.6	64.1	16.3	19.2	64.3	16.7
Switzerland	11.6	10.9	75.2	78.6	109.2	108.6	1.5	1.5	19.7	66.4	13.8	17.2	68.1	14.7
Ukraine	14.9	9.7	69.3	68.8	114.8	116.4	2.0	1.4	21.4	66.7	11.9	17.8	67.9	14.3
United Kingdom	12.4	11.9	72.8	77.1	109.0	107.1	1.7	1.7	20.9	64.0	15.1	18.9	65.5	15.8
Yugoslavia	18.4	12.6	70.3	72.5	106.5	107.9	2.4	1.8	24.1	66.1	9.8	19.8	66.9	13.3
OCEANIA	**20.9**	**18.4**	**68.2**	**73.9**	**108.7**	**106.9**	**2.8**	**2.5**	**29.3**	**62.7**	**8.0**	**25.5**	**64.9**	**9.7**
Australia	16.0	14.3	73.4	78.3	109.8	107.7	2.1	1.9	25.3	65.1	9.6	21.0	67.1	11.9
Fiji	33.2	22.6	67.1	72.6	105.3	106.1	4.0	2.8	39.1	58.0	2.8	31.3	64.3	4.5
New Zealand	17.2	15.4	72.4	77.2	109.2	106.7	2.2	2.0	26.7	63.3	10.0	23.0	65.6	11.3
Papua New Guinea	39.6	32.3	49.7	57.9	101.0	102.6	5.9	4.7	43.0	55.5	1.6	38.7	58.3	3.0
Solomon Islands	45.0	36.1	65.3	71.6	106.2	106.2	7.1	5.0	47.6	49.3	3.1	43.0	54.1	2.9
DEVELOPING COUNTRIES	**32.8**	**25.4**	**56.7**	**63.6**	**103.8**	**105.0**	**4.7**	**3.1**	**39.3**	**56.6**	**4.1**	**32.8**	**62.2**	**5.0**
DEVELOPED COUNTRIES	**14.9**	**11.4**	**72.2**	**74.5**	**110.8**	**111.0**	**1.9**	**1.6**	**22.5**	**65.9**	**11.6**	**18.3**	**67.5**	**14.2**

Source: United Nations Population Division, *World Resources 1998–99*.

Table F

World Countries: Mortality, Health, and Nutrition, 1970–1995

COUNTRY	MORTALITY — Crude Death Rate per 1,000 of the Population			Infant Mortality Rate per 1,000 Live Births			Under-5 Mortality Rate per 1,000 Live Births			HEALTH — Population per Physician		Population per Nursing Person		Total Expenditure on Health as a % of GDP	NUTRITION — Calories Available as a Percent of Need
	1975–80	1995–00	2015–20	1975–80	1995–00	2015–20	1960	1980	1995	1970	1994	1970	1994	1995	1995
AFRICA	17.7	12.9	8.4	120	85	55	X	X	X	X	X	X	X	X	X
Algeria	13.4	5.6	4.6	112	44	21	243	134	61	8100	X	X	X	7.0	123
Angola	24.4	18.7	11.1	160	124	79	345	261	292	X	X	X	X	X	80
Benin	21.1	12.4	6.9	122	84	46	210	176	142	28,960	X	2,610	X	4.3	104
Botswana	11.3	13.0	7.5	76	56	33	170	94	52	15,540	X	1,920	X	X	97
Burkina Faso	21.9	17.7	10.7	127	97	63	318	218	164	95,690	X	X	X	8.5	94
Burundi	18.8	17.0	10.8	127	114	78	255	193	176	58,570	17,240	6,910	4,800	3.3	84
Cameroon	17.6	11.9	6.5	102	58	34	264	173	106	29,390	12,000	2,610	2,000	2.6	95
Central African Republic	20.6	16.4	10.3	122	96	60	294	202	165	44,020	X	2,480	X	4.2	82
Chad	23.1	17.3	11.3	154	115	79	325	254	152	61,900	29,410	8,020	X	6.3	73
Congo (Zaire)	17.5	13.5	8.0	117	89	53	286	204	185	X	X	X	X	2.4	96
Congo	17.4	14.6	8.5	91	90	55	220	125	108	9,940	X	810	X	X	103
Cote d'Ivoire	17.5	13.8	8.7	117	86	54	300	180	150	15,540	X	1,930	X	3.3	111
Egypt	14.2	7.1	6.0	131	54	25	258	180	51	2,030	1,340	2,480	500	2.6	132
Equatorial Guinea	22.7	16.2	10.7	149	107	73	316	243	X	X	X	X	X	X	X
Eritrea	18.9	14.7	9.0	130	98	57	X	X	195	X	X	X	X	X	X
Ethiopia	22.1	16.2	9.4	149	107	65	260	260	195	85,690	X	X	X	3.6	73
Gabon	19.2	14.3	9.3	122	85	47	194	194	148	5,520	X	570	X	X	104
Gambia	24.9	17.4	12.3	167	122	85	276	X	110	24,420	X	X	X	X	X
Ghana	15.3	10.4	6.9	103	73	43	157	170	130	12,910	X	690	X	3.5	93
Guinea	25.4	18.4	12.1	167	124	87	337	278	219	50,650	X	3,370	X	3.9	97
Guinea-Bissau	26.2	20.6	14.0	176	132	94	336	290	227	17,500	X	2,860	X	X	97
Kenya	15.5	11.3	6.2	88	65	37	202	112	90	8,000	X	2,520	X	4.3	89
Lesotho	16.5	10.6	6.6	121	72	37	204	173	154	30,400	X	X	X	X	93
Liberia	18.1	15.3	8.0	167	153	67	288	235	216	X	X	3,860	X	X	98
Libya	12.7	6.9	4.5	107	56	25	269	150	63	X	X	X	X	X	140
Madagascar	16.1	9.9	6.0	150	77	42	364	216	164	10,310	X	2,500	X	2.6	95
Malawi	24.0	22.4	12.6	177	142	93	365	290	219	76,580	50,360	5,330	1,980	5.0	88
Mali	24.1	17.1	11.1	191	149	107	400	210	210	45,320	21,180	2,670	2,050	5.2	96
Mauritania	20.0	13.1	8.8	125	92	59	321	249	195	17,960	X	3,750	X	X	106
Mauritius	6.3	6.5	6.9	38	15	8	84	42	23	4,170	X	610	X	X	128
Morocco	13.0	6.7	5.8	110	51	24	215	145	75	13,090	X	X	X	2.6	125
Mozambique	20.8	17.5	11.1	160	110	74	331	269	275	18,870	X	4,280	X	5.9	77
Namibia	15.1	11.8	7.4	98	60	36	206	114	78	X	4,320	X	X	X	X
Niger	23.8	17.1	10.9	157	114	79	320	320	320	60,360	35,140	5,690	660	5.0	95
Nigeria	19.2	13.9	9.2	105	77	51	204	196	191	20,530	X	4,370	X	2.7	93
Rwanda	20.2	19.7	11.7	133	125	84	191	222	139	60,130	X	5,630	X	2.4	82
Senegal	21.7	14.5	9.6	97	62	39	303	221	110	15,810	X	1,670	X	1.5	98
Sierra Leone	29.0	25.7	16.1	192	169	114	385	301	284	17,830	X	2,700	X	5.6	83
Somalia	22.7	16.9	10.7	149	112	77	294	246	211	X	X	X	X	3.3	81
South Africa	12.4	7.9	6.0	72	48	27	91	91	67	X	X	300	X	X	126
Sudan	17.8	11.7	8.4	97	71	45	292	210	115	X	X	X	X	X	87

(Continued on next page)

COUNTRY	MORTALITY									HEALTH					NUTRITION
	Crude Death Rate per 1,000 of the Population			Infant Mortality Rate per 1,000 Live Births			Under-5 Mortality Rate per 1,000 Live Births			Population per Physician		Population per Nursing Person		Total Expenditure on Health as a % of GDP	Calories Available as a Percent of Need
	1975–80	1995–00	2015–20	1975–80	1995–00	2015–20	1960	1980	1995	1970	1994	1970	1994	1995	1995
Swaziland	16.0	9.2	5.7	108	65	34	233	151	X	X	X	X	X	X	X
Tanzania	16.4	13.5	8.0	113	80	51	249	202	160	22,900	3,400	X	X	4.7	X
Togo	17.4	14.9	8.5	117	86	52	164	175	128	28,860	X	1,590	X	4.1	99
Tunisia	10.0	5.9	5.5	88	37	18	244	102	37	5,930	1,540	940	300	4.9	131
Uganda	17.6	21.0	10.7	114	113	72	218	181	185	9,210	X	X	X	3.4	93
Zambia	16.5	18.0	9.3	94	103	55	220	160	203	13,640	11,430	1,730	610	3.2	87
Zimbabwe	13.1	14.6	8.0	86	68	39	181	125	74	6,310	X	650	X	6.2	94
NORTH AMERICA	**8.5**	**8.6**	**8.8**	**14**	**7**	**5**	**X**	**13**	**8**	**X**	**X**	**X**	**X**	**X**	**X**
Canada	7.2	7.4	8.7	12	6	5	33	13	8	680	X	140	X	9.9	122
United States	8.6	8.7	8.8	14	7	5	30	15	10	630	X	160	X	13.3	138
CENTRAL AMERICA	**8.3**	**5.3**	**5.5**	**62**	**33**	**22**	**X**	**X**	**X**	**X**	**X**	**X**	**X**	**X**	**X**
Belize	6.5	4.3	3.6	45	30	21	X	X	X	X	X	X	X	X	X
Costa Rica	4.8	3.8	4.7	30	12	8	112	29	16	1,620	X	460	X	X	121
Cuba	6.0	7.0	8.9		22	9	650	26	10	X	X	X	X	X	135
Dominican Republic	8.4	5.3	5.6	84	34	19	152	94	44	X	X	1,400	X	3.7	102
El Salvador	11.2	6.0	5.6	87	39	25	210	120	40	4,100	X	890	X	5.9	102
Guatemala	12.0	6.7	5.2	82	40	25	205	136	67	3,660	X	X	X	3.7	103
Haiti	16.0	12.8	8.8	121	82	51	270	195	124	X	X	X	X	7.0	89
Honduras	11.0	5.4	4.8	81	35	21	203	100	38	2,630	X	X	X	4.5	96
Jamaica	7.4	5.8	5.3	26	12	7	76	39	13	1,480	X	530	X	X	114
Mexico	7.7	5.1	5.6	57	31	21	141	81	32	2,150	X	1,620	X	3.2	X
Nicaragua	11.3	5.8	4.8	90	44	28	209	143	60	X	1,490	X	X	8.6	99
Panama	6.3	5.1	5.8	35	21	13	104	31	20	1,630	X	1,540	X	X	98
Trinidad and Tobago	7.1	6.0	6.4	32	14	7	73	40	18	2,250	X	150	X	X	114
SOUTH AMERICA	**8.9**	**6.8**	**6.8**	**72**	**36**	**23**	**X**	**X**	**X**	**X**	**X**	**960**	**X**	**X**	**X**
Argentina	8.9	7.9	7.5	39	22	14	68	41	27	530	X	X	X	4.2	131
Bolivia	16.0	9.1	6.4	131	66	33	252	170	105	1,970	X	2,990	X	4.0	84
Brazil	9.1	7.1	7.4	79	42	26	181	93	60	2,030	X	4,140	X	4.2	114
Chile	7.5	5.6	6.6	45	13	9	138	35	15	2,160	2,150	460	330	4.7	102
Colombia	7.6	5.6	5.9	59	24	21	132	59	36	2,260	960	X	X	4.0	106
Ecuador	9.8	6.0	6.0	82	46	29	180	101	40	2,870	X	2,640	600	4.1	105
Guyana	9.2	7.4	7.0	67	58	37	128	86	X	X	X	X	X	X	X
Paraguay	7.8	5.4	4.7	51	39	27	90	61	34	2,300	1,260	2,210	X	2.8	116
Peru	10.9	6.4	6.1	99	45	24	236	130	55	1,920	940	X	X	3.2	87
Suriname	7.3	5.5	5.4	44	24	12	96	52	X	X	X	X	X	X	X
Uruguay	10.1	10.4	10.3	42	17	14	47	42	21	910	X	X	X	4.6	101
Venezuela	5.8	4.7	5.2	39	21	14	70	42	24	1,130	640	450	330	3.6	99
ASIA	**10.5**	**7.9**	**7.4**	**94**	**56**	**32**	**X**	**X**	**X**	**X**	**X**	**X**	**X**	**X**	**X**
Afghanistan	24.0	20.8	12.8	183	154	118	360	280	257	X	X	X	X	X	72
Armenia	5.9	7.5	9.0	22	25	19	X	X	31	350	260	140	100	X	X
Azerbaijan	7.0	6.6	7.3	41	33	24	X	X	50	390	260	130	110	X	X
Bangladesh	18.9	9.7	6.7	137	78	38	247	211	115	8,450	5,220	65,810	11,350	3.2	88
Bhutan	21.2	13.7	8.0	165	104	56	324	249	189	X	X	X	X	X	128

(Continued on next page)

COUNTRY	MORTALITY — Crude Death Rate per 1,000 of the Population			MORTALITY — Infant Mortality Rate per 1,000 Live Births			MORTALITY — Under-5 Mortality Rate per 1,000 Live Births			HEALTH — Population per Physician		HEALTH — Population per Nursing Person		HEALTH — Total Expenditure on Health as a % of GDP	NUTRITION — Calories Available as a Percent of Need
	1975–80	1995–00	2015–20	1975–80	1995–00	2015–20	1960	1980	1995	1970	1994	1970	1994	1995	1995
Cambodia	40.0	12.2	8.1	263	102	55	217	330	174	X	X	X	X	X	96
China	6.7	7.1	8.2	52	38	20	209	65	47	1,500	1,060	2,500	1,490	3.5	112
Georgia	8.7	9.5	10.7	36	23	18	X	X	26	280	180	110	80	X	X
India	13.9	9.0	7.6	129	72	45	236	177	115	4,950	X	3,760	X	6.0	101
Indonesia	15.1	7.6	7.0	105	48	23	216	128	75	27,440	X	4,910	X	2.0	121
Iran	11.8	6.0	4.4	100	39	20	233	126	40	3,270	X	1,780	X	2.6	125
Iraq	8.8	8.5	4.5	84	95	23	171	83	71	X	X	X	X	X	128
Israel	6.8	6.4	6.5	18	7	5	39	19	9	410	X	X	X	4.2	125
Japan	6.1	8.2	12.1	9	4	4	40	11	6	890	X	310	X	6.8	125
Jordan	9.6	4.8	3.7	65	30	15	149	66	25	2,480	770	870	500	3.8	110
Kazakhstan	8.8	8.5	8.7	45	34	24	X	X	47	460	250	120	90	X	X
Korea, North	6.2	5.5	6.7	38	22	13	120	43	30	X	X	X	X	X	121
Korea, South	7.1	6.4	8.4	30	9	6	124	18	9	2,220	950	1,190	450	6.6	120
Kuwait	4.2	2.2	4.0	34	14	7	128	35	14	1,050	310	260	110	X	X
Kyrgyzstan	9.6	7.4	6.5	55	39	27	X	X	54	480	310	140	X	X	X
Laos	20.7	13.5	7.5	135	86	49	233	190	134	15,160	4,450	1,380	490	2.5	111
Lebanon	8.7	6.4	6.0	48	29	17	91	52	40	X	X	X	X	X	127
Malaysia	7.2	4.8	5.1	34	11	7	105	42	13	4,310	2,410	1,270	470	3.0	120
Mongolia	11.4	6.8	5.2	88	52	33	185	112	74	580	360	250	X	X	97
Myanmar (Burma)	14.9	9.8	7.1	114	78	33	237	146	150	8,820	12,900	3,050	1,240	X	114
Nepal	18.4	11.0	6.6	142	82	40	279	177	114	52,050	16,110	17,970	2,300	4.5	100
Oman	12.4	4.3	3.2	95	25	12	300	95	25	9,270	X	3,820	X	X	X
Pakistan	15.4	7.8	5.2	130	74	45	221	151	137	4,670	X	7,020	X	3.4	99
Philippines	9.0	5.7	5.3	62	35	18	102	70	53	9,270	X	2,680	X	2.0	104
Saudi Arabia	10.7	4.2	3.7	75	23	10	292	90	34	7,460	710	2,080	460	4.8	120
Singapore	5.1	5.0	7.1	13	5	5	40	13	6	1,520	X	280	X	1.9	136
Sri Lanka	7.1	5.9	6.6	44	15	8	130	52	19	5,900	X	1,290	X	3.7	101
Syria	8.9	4.9	4.0	67	33	17	201	73	36	X	X	X	X	2.1	126
Tajikistan	8.9	6.9	5.5	69	56	38	X	X	79	630	430	190	140	X	X
Thailand	8.3	6.6	7.4	56	30	13	146	61	32	8,290	4,420	1,170	910	5.0	103
Turkey	10.2	6.5	6.7	120	44	22	217	141	50	2,230	980	1,000	1,110	4.0	127
Turkmenistan	9.6	7.6	6.0	73	57	38	X	X	85	460	280	140	90	X	X
United Arab Emirates	7.4	2.9	6.0	38	15	6	240	64	19	1,120	1,100	X	580	X	X
Uzbekistan	8.5	6.6	5.7	58	43	30	X	X	62	490	280	150	90	X	X
Viet Nam	11.4	7.0	5.8	82	37	21	219	105	14	X	2,300	4,310	400	2.1	103
Yemen	20.7	10.4	5.2	158	80	38	X	X	110	34,790	X	X	X	X	X
EUROPE	**10.5**	**11.5**	**12.4**	**22**	**12**	**8**	**X**	**X**	**X**	**X**	**X**	**X**	**X**	**X**	**X**
Albania	6.4	6.2	7.1	50	32	20	151	57	40	1,070	X	230	X	X	107
Austria	12.4	10.0	10.7	17	6	5	43	17	7	550	230	300	X	8.5	133
Belarus	9.4	12.1	13.3	22	15	10	X	X	20	390	230	120	90	X	X
Belgium	11.7	10.5	11.3	14	7	5	35	15	10	640	X	X	X	8.1	149
Bosnia-Herzegovina	6.5	7.3	10.9	36	13	9	X	X	17	X	X	X	X	X	X
Bulgaria	10.5	13.3	13.9	22	16	10	70	25	19	540	X	240	X	5.4	148
Croatia	10.8	11.7	13.4	21	10	7	X	X	14	X	X	X	X	X	X
Czech Republic	13.2	12.0	12.3	18	9	6	X	X	10	X	270	X	X	X	X
Denmark	10.4	11.8	11.8	9	7	5	25	10	7	690	X	X	X	7.0	135
Estonia	12.1	13.1	13.9	22	12	6	X	X	22	300	260	110	130	X	X

(Continued on next page)

COUNTRY	MORTALITY									HEALTH					NUTRITION
	Crude Death Rate per 1,000 of the Population			Infant Mortality Rate per 1,000 Live Births			Under-5 Mortality Rate per 1,000 Live Births			Population per Physician		Population per Nursing Person		Total Expenditure on Health as a % of GDP	Calories Available as a Percent of Need
	1975–80	1995–00	2015–20	1975–80	1995–00	2015–20	1960	1980	1995	1970	1994	1970	1994	1995	1995
Finland	9.3	10.2	11.3	9	5	5	28	9	5	960	X	130	X	8.9	113
France	10.3	9.1	10.5	11	7	6	34	13	9	750	X	270	X	9.1	143
Germany	12.2	10.9	12.6	15	6	5	40	16	7	580	X	X	X	9.1	X
Greece	8.8	9.9	12.3	25	8	7	64	23	10	620	X	990	X	4.8	151
Hungary	12.8	14.6	14.5	27	14	9	57	26	14	510	X	210	X	6.0	137
Iceland	6.4	6.8	7.4	9	5	5	22	9	X	X	X	X	X	8.3	X
Ireland	10.2	8.6	8.9	15	6	5	36	14	7	980	X	160	X	8.0	157
Italy	9.8	9.9	12.4	18	7	6	50	17	8	550	X	X	X	8.3	139
Latvia	12.4	13.8	14.6	23	16	11	X	X	26	X	280	X	120	X	X
Lithuania	9.9	11.6	12.9	22	13	7	X	X	19	360	230	130	90	X	X
Macedonia	7.1	7.4	9.3	57	23	13	X	X	31	X	430	X	X	X	X
Moldova	10.6	10.8	10.6	46	26	19	X	X	34	490	250	X	X	X	X
Netherlands	8.1	8.8	10.3	10	6	5	22	11	8	800	X	300	X	8.7	114
Norway	10.0	10.7	10.3	9	5	5	23	11	8	720	X	160	X	8.4	120
Poland	9.2	10.7	11.1	23	13	8	70	24	16	700	450	260	190	5.1	131
Portugal	10.3	10.9	11.7	30	8	5	112	31	11	1,110	X	860	X	6.2	136
Romania	9.7	11.6	12.8	31	24	14	82	36	29	840	540	430	X	3.9	116
Russian Federation	10.3	14.5	14.8	30	19	12	X	X	30	340	220	110	90	X	X
Slovak Republic	10.4	10.5	10.8	22	12	8	X	X	15	X	290	X	110	X	X
Slovenia	10.4	10.7	13.3	17	7	5	X	X	8	X	X	X	X	6.5	X
Spain	8.1	9.4	11.3	16	7	6	57	16	9	750	X	140	X	6.5	141
Sweden	10.9	11.2	10.4	8	5	5	20	9	5	730	X	140	X	8.8	111
Switzerland	9.0	9.1	10.4	10	5	5	27	11	7	700	X	X	X	8.0	136
Ukraine	11.0	13.8	14.4	23	18	10	X	X	24	360	220	110	90	X	X
United Kingdom	11.9	11.1	11.0	14	6	5	27	14	7	810	X	240	X	6.6	130
Yugoslavia (Serbia-Montenegro)	9.2	9.7	11.6	38	19	12	X	X	23	X	X	X	X	X	X
OCEANIA	**8.7**	**7.7**	**7.6**	**35**	**24**	**15**	**X**	**X**	**X**	**X**	**X**	**X**	**X**	**X**	**X**
Australia	7.7	7.5	8.1	13	6	5	24	13	8	840	X	X	X	8.6	124
Fiji	5.8	4.6	5.5	37	20	10	97	42	X	X	X	X	X	X	X
New Zealand	8.2	7.9	8.0	14	7	7	26	16	9	870	X	150	X	7.7	131
Papua New Guinea	15.0	9.9	7.0	77	61	37	248	95	95	11,640	12,750	1,710	1,160	4.4	114
Solomon Islands	7.4	4.1	3.4	47	23	12	X	X	X	X	X	X	X	X	X
DEVELOPING COUNTRIES	**11.5**	**8.5**	**7.4**	**98**	**6**	**2**	**38**	**X**	**X**	**X**	**X**	**X**	**X**	**X**	**X**
DEVELOPED COUNTRIES	**9.5**	**10.3**	**11.2**	**18**	**9**	**7**	**X**	**X**	**X**	**X**	**X**	**X**	**X**	**X**	**X**

Sources: United Nations Children's Fund, United Nations Development Programme, UNESCO; World Resources Institute, World Resources 1998–99, World Almanac and Book of Facts 1998.

(Continued on next page)

Table G

World Countries: Illiteracy and Education

COUNTRY	ILLITERACY				EDUCATION							FEMALE ENROLLMENT as a % of the Total			
	Adult Male Illiteracy (% of above age 15)		Adult Female Illiteracy (% of above age 15)		Preprimary	Primary % of Relevant Age Group		Secondary % of Relevant Age Group		Higher Ed. % of Relevant Age Group		Primary		Secondary	
	1970	1995	1970	1995	1995	1980	1995	1980	1995	1980	1995	1980	1995	1980	1995
AFRICA															
Algeria	61	26	88	51	2	94	107	33	62	6	11	42	46	59	54
Angola	84	X	93	X	68	174	88	21	14	0	1	47	X	33	X
Benin	83	51	96	74	3	67	72	16	16	1	3	32	X	26	X
Botswana	40	20	69	40	X	91	115	19	56	1	4	55	50	56	53
Burkina Faso	87	71	98	91	X	17	38	3	8	0	1	37	39	33	34
Burundi	70	51	93	78	X	26	70	3	7	1	1	39	45	25	38
Cameroon	53	25	82	48	11	98	88	18	27	2	X	45	47	34	40
Central African Republic	79	32	95	48	71	58	14	10	1	1	X	37	X	25	X
Chad	64	38	88	65	1	X	55	X	9	X	1	X	32	X	17
Congo (Zaire)	39	13	72	32	1	80	72	24	26	1	2	42	43	X	30
Congo	51	17	79	33	1	141	114	74	53	5	X	48	48	40	41
Cote d'Ivoire	75	50	94	70	2	75	69	19	23	3	4	40	43	28	34
Egypt	53	36	82	61	9	73	100	50	74	16	18	40	45	36	44
Equatorial Guinea	35	X	71	X	X	X	X	X	X	X	X	X	X	X	X
Eritrea	X	X	X	X	X	X	X	X	X	X	X	X	44	X	42
Ethiopia	76	55	91	75	1	36	31	9	11	0	1	35	38	36	46
Gabon	61	26	86	47	X	X	142	X	X	X	X	49	50	42	X
Gambia	72	47	93	75	24	53	73	11	22	X	2	35	41	30	X
Ghana	55	24	83	47	X	79	76	41	37	2	X	44	46	38	X
Guinea	75	50	94	78	9	36	48	17	12	5	X	33	33	28	24
Guinea-Bissau	57	32	83	58	X	68	64	6	X	X	X	32	X	22	X
Kenya	41	14	73	30	36	115	85	20	24	1	X	47	49	42	44
Lesotho	38	19	66	38	X	102	99	18	28	1	2	59	53	60	59
Liberia	73	X	93	X	X	X	X	X	X	X	X	X	X	X	X
Libya	43	12	89	37	X	125	110	76	97	8	16	47	49	39	X
Madagascar	34	X	57	X	X	133	72	X	14	3	3	49	49	X	50
Malawi	42	28	80	58	X	60	135	3	98	1	2	41	47	29	39
Mali	89	61	96	77	3	26	34	8	9	1	X	36	39	29	34
Mauritania	63	50	83	74	0	37	78	11	15	X	4	35	45	21	36
Mauritius	24	13	45	21	85	93	107	50	62	1	6	49	49	48	51
Morocco	67	43	91	69	63	83	83	26	39	6	11	37	42	38	42
Mozambique	74	42	94	77	X	99	60	5	7	0	1	43	42	29	40
Namibia	X	X	X	X	11	X	133	X	62	X	8	X	50	X	55
Niger	89	79	99	93	1	25	29	5	7	0	X	35	38	29	33
Nigeria	68	33	89	53	X	105	89	16	30	2	X	43	44	36	X
Rwanda	55	30	82	48	X	63	82	3	11	0	X	48	50	36	X
Senegal	76	57	93	77	2	46	65	11	16	3	3	40	43	34	35
Sierra Leone	78	55	95	82	X	52	X	14	X	1	X	42	X	30	X

(Continued on next page)

(Continued on next page)

COUNTRY	ILLITERACY				EDUCATION								FEMALE ENROLLMENT as a % of the Total			
	ADULT MALE ILLITERACY (% of above age 15)		ADULT FEMALE ILLITERACY (% of above age 15)		Preprimary	Primary % of Relevant Age Group		Secondary % of Relevant Age Group		Higher Ed. % of Relevant Age Group			Primary		Secondary	
	1970	1995	1970	1995	1995	1980	1995	1980	1995	1980	1995		1980	1995	1980	1995
Somalia	95	X	99	X	X	X	X	X	X	X	X		X	X	X	X
South Africa	28	18	32	18	28	85	117	X	84	X	17		X	49	X	53
Sudan	67	42	91	65	37	50	54	16	13	2	X		40	44	37	X
Swaziland	47	X	56	X	X	X	X	X	X	X	X		X	X	X	X
Tanzania	45	21	79	43	0	93	67	3	5	X	1		47	49	33	43
Togo	64	33	90	63	3	118	118	33	27	2	3		38	40	24	26
Tunisia	58	21	83	45	X	102	116	27	61	5	13		42	47	39	47
Uganda	47	26	78	50	X	50	73	5	12	1	2		43	44	29	38
Zambia	36	14	68	29	X	90	89	16	28	2	3		47	48	35	38
Zimbabwe	25	10	62	20	X	85	116	8	47	1	7		48	48	42	44
NORTH AND CENTRAL AMERICA																
Belize	X	X	X	X	X	X	X	X	X	X	X		X	X	X	X
Canada	X	X	X	X	63	99	102	88	106	57	103		49	48	49	49
Costa Rica	12	5	13	5	70	105	107	48	50	21	32		49	49	54	52
Cuba	15	4	22	5	89	106	105	81	80	17	14		48	49	51	54
Dominican Republic	30	18	34	18	20	118	103	42	41	X	X		40	50	X	57
El Salvador	39	27	48	30	31	74	88	25	32	13	18		49	49	43	48
Guatemala	49	38	64	51	32	71	84	18	25	8	8		45	46	43	X
Haiti	72	52	80	58	X	76	X	14	X	1	X		46	X	47	X
Honduras	43	27	49	27	14	98	111	30	32	8	10		50	50	50	X
Jamaica	34	19	27	11	81	103	109	67	66	7	6		50	49	52	X
Mexico	20	8	30	13	71	120	115	49	58	14	14		49	48	43	48
Nicaragua	42	35	44	33	20	98	110	43	47	13	9		51	50	52	53
Panama	20	9	22	10	76	106	106	61	68	21	30		48	X	51	X
Trinidad and Tobago	5	1	11	3	10	99	96	70	72	4	8		50	49	50	51
United States	1	X	1	X	68	99	102	91	97	56	81		49	X	49	49
SOUTH AMERICA																
Argentina	7	4	8	4	50	106	108	56	72	22	38		49	X	64	X
Bolivia	29	10	54	24	X	87	X	37	45	16	X		47	X	X	X
Brazil	28	17	35	27	56	98	112	33	45	11	11		49	X	X	X
Chile	11	5	13	5	96	109	99	53	69	12	28		49	49	55	54
Colombia	18	9	21	9	28	124	114	41	67	9	17		50	50	50	X
Ecuador	21	8	30	12	49	117	109	53	50	35	X		49	49	48	47
Guyana	6	X	12	X	X	X	X	X	X	X	X		X	X	X	X
Paraguay	14	7	24	9	38	106	109	27	38	9	10		X	X	X	X
Peru	17	6	40	17	36	114	123	59	70	17	31		X	X	X	x
Suriname	12	X	24	X	X	X	X	X	X	X	X		X	X	X	X
Uruguay	8	3	7	2	33	107	111	62	82	17	27		49	49	58	55
Venezuela	20	8	27	10	43	93	94	21	35	21	29		50	50	X	58

COUNTRY	ADULT MALE ILLITERACY (% of above age 15)		ADULT FEMALE ILLITERACY (% of above age 15)		Preprimary	Primary % of Relevant Age Group		Secondary % of Relevant Age Group		Higher Ed. % of Relevant Age Group		Female Enrollment Primary		Female Enrollment Secondary	
	1970	1995	1970	1995	1995	1980	1995	1980	1995	1980	1995	1980	1995	1980	1995
ASIA															
Afghanistan	75	X	97	X	X	X	X	X	X	X	X	X	X	X	X
Armenia	X	X	X	X	22	X	82	X	79	30	49	X	50	X	X
Azerbaijan	X	X	X	X	20	115	104	93	74	24	20	X	47	X	48
Bangladesh	64	51	88	74	X	61	92	18	X	3	X	37	X	24	X
Bhutan	69	X	91	X	X	X	X	X	X	X	X	X	X	X	X
Cambodia	X	20	77	47	5	X	122	32	27	1	2	X	44	X	38
China	33	10	64	27	29	113	118	46	67	2	5	45	47	40	44
Georgia	X	X	X	X	32	X	82	X	73	30	38	X	X	X	X
India	52	35	82	62	5	83	100	30	49	5	6	39	43	32	38
Indonesia	31	10	56	22	19	107	114	29	48	4	11	46	48	36	46
Iran	X	22	X	34	7	87	99	42	69	X	15	40	47	39	45
Iraq	56	29	85	55	8	113	90	57	44	9	X	46	45	32	39
Israel	X	X	X	X	71	95	99	73	89	29	41	49	49	56	53
Japan	1	X	1	X	49	101	102	93	99	31	40	49	49	50	50
Jordan	27	7	66	21	25	104	94	75	X	27	X	48	49	46	55
Kazakhstan	X	X	X	X	29	84	96	93	83	34	33	X	49	X	52
Korea, North	X	X	X	X	X	X	X	X	X	X	X	X	X	X	X
Korea, South	6	1	20	3	85	110	101	78	101	15	52	49	48	46	47
Kuwait	35	18	55	25	52	102	73	80	64	11	25	48	49	46	49
Kyrgyzstan	X	X	X	X	8	116	107	110	81	16	14	49	50	49	51
Laos	55	31	82	56	7	113	107	21	25	0	2	45	43	38	39
Lebanon	13	10	27	20	74	111	109	59	81	30	27	48	48	51	53
Malaysia	31	11	56	22	X	93	91	48	61	4	11	49	49	48	51
Mongolia	25	X	48	X	23	107	88	91	59	X	15	49	51	52	58
Myanmar (Burma)	16	11	40	22	X	91	100	22	32	5	5	48	48	45	X
Nepal	76	59	96	86	X	86	110	22	38	3	5	28	39	X	X
Oman	X	X	X	X	3	51	80	12	66	X	5	34	48	25	48
Pakistan	69	50	91	76	X	39	74	14	26	X	3	33	31	26	X
Philippines	14	5	17	6	13	112	116	64	79	24	27	49	50	53	X
Saudi Arabia	51	29	79	50	8	61	78	29	58	7	15	39	47	37	44
Singapore	13	4	39	14	X	108	104	58	62	8	34	48	X	51	X
Sri Lanka	12	7	29	13	X	103	113	55	75	3	5	48	48	51	51
Syria	39	14	79	44	7	100	101	46	44	17	18	43	47	37	44
Tajikistan	X	X	X	X	10	X	89	X	82	24	20	X	49	X	47
Thailand	14	4	30	8	58	99	87	29	55	15	20	48	49	46	50
Turkey	27	8	60	28	6	96	105	35	56	5	18	45	47	35	39
Turkmenistan	X	X	X	X	X	X	X	X	X	23	X	X	X	X	X
United Arab Emirates	39	21	60	20	57	89	95	52	78	3	9	48	48	45	51
Uzbekistan	X	X	X	X	54	81	77	105	93	29	32	49	49	46	48
Viet Nam	18	X	35	X	35	109	114	42	47	2	4	47	X	47	X
Yemen	X	X	X	X	1	X	79	X	23	X	4	X	28	X	X

(Continued on next page)

COUNTRY	ILLITERACY				EDUCATION							FEMALE ENROLLMENT as a % of the Total			
	ADULT MALE ILLITERACY (% of above age 15)		ADULT FEMALE ILLITERACY (% of above age 15)		Preprimary	Primary % of Relevant Age Group		Secondary % of Relevant Age Group		Higher Ed. % of Relevant Age Group		Primary		Secondary	
	1970	1995	1970	1995	1995	1980	1995	1980	1995	1980	1995	1980	1995	1980	1995
EUROPE															
Albania	X	X	X	X	38	113	87	67	35	8	19	47	48	59	54
Austria	X	X	X	X	76	99	101	99	104	26	45	49	49	49	49
Belarus	X	X	X	X	80	104	97	98	X	39	X	X	48	X	50
Belgium	1	X	1	X	116	104	103	91	144	26	49	49	49	X	X
Bosnia-Herzegovina	X	X	X	X	X	X	X	X	X	X	X	X	X	X	X
Bulgaria	6	X	X	X	62	98	94	84	78	16	39	49	48	68	67
Croatia	X	X	X	X	31	X	86	X	82	19	28	49	49	X	65
Czech Republic	X	X	X	X	91	X	96	X	96	18	21	X	49	X	52
Denmark	X	X	X	X	81	996	99	105	118	28	45	49	49	51	52
Estonia	X	X	X	X	56	98	91	X	86	25	38	49	48	X	53
Finland	X	X	X	X	39	96	100	100	116	32	67	49	49	53	53
France	1	X	2	X	84	111	106	85	111	25	50	48	48	49	51
Germany	X	X	X	X	84	X	102	98	103	34	43	49	49	X	50
Greece	7	X	24	X	61	103	X	81	95	17	38	48	48	50	50
Hungary	2	X	2	X	86	96	97	70	81	14	19	49	49	65	63
Iceland	X	X	X	X	X	X	X	X	X	X	X	X	X	X	X
Ireland	1	X	1	X	107	100	104	90	114	18	37	49	49	51	50
Italy	X	X	1	X	96	100	98	72	74	27	41	49	49	48	50
Latvia	X	X	X	X	44	78	89	100	85	24	26	49	48	X	52
Lithuania	X	X	X	X	36	79	96	114	84	35	28	49	48	X	52
Macedonia	X	X	X	X	24	100	89	61	57	28	18	X	48	X	60
Moldova	X	X	X	X	45	83	94	78	80	30	25	49	49	51	51
Netherlands	X	X	X	X	100	100	107	93	139	29	49	49	50	52	52
Norway	X	X	X	X	98	100	99	94	92	26	55	49	49	51	51
Poland	1	X	3	X	45	100	98	77	96	18	27	49	48	51	51
Portugal	22	X	35	X	58	123	128	37	102	11	34	48	X	71	68
Romania	4	X	X	X	53	102	100	71	66	12	18	49	49	65	53
Russian Federation	X	X	X	X	63	102	108	96	87	46	43	49	49	51	52
Slovak Republic	X	X	X	X	71	X	97	X	91	X	20	X	49	X	51
Slovenia	X	X	X	X	66	X	98	X	91	X	32	X	49	X	52
Spain	6	X	14	X	69	109	105	87	118	23	46	49	48	51	51
Sweden	X	X	X	X	60	97	105	88	132	31	43	49	49	51	51
Switzerland	X	X	X	X	94	X	107	X	91	18	32	49	49	49	50
Ukraine	X	X	X	X	54	102	87	94	91	42	41	49	49	X	51
United Kingdom	X	X	X	X	29	103	115	83	134	19	48	49	49	49	49
Yugoslavia (Serbia-Montenegro)	X	X	X	X	31	29	72	X	65	X	21	49	49	51	51
OCEANIA															
Australia	X	X	X	X	73	112	108	71	147	25	72	49	49	50	49
Fiji	21	X	33	X	X	X	X	X	X	X	X	X	X	X	X

(Continued on next page)

COUNTRY	ILLITERACY				EDUCATION										
	ADULT MALE ILLITERACY (% of above age 15)		ADULT FEMALE ILLITERACY (% of above age 15)		Preprimary	Primary % of Relevant Age Group		Secondary % of Relevant Age Group		Higher Ed. % of Relevant Age Group		FEMALE ENROLLMENT as a % of the Total			
												Primary		Secondary	
	1970	1995	1970	1995	1995	1980	1995	1980	1995	1980	1995	1980	1995	1980	1995
New Zealand	X	X	X	X	77	111	104	83	117	27	58	49	49	49	50
Papua New Guinea	40	19	67	37	1	59	80	12	14	2	3	41	45	32	41
Solomon Islands	X	X	X	X	X	X	X	X	X	X	X	X	X	X	X

Sources: World Development Indicators (World Bank, 1998); *Human Development Report* (World Bank, 1995); *World Resources 1998-99* (World Resources Institute, 1996); United Nations Children's Fund, UNESCO, the World Health Organization.

Table H

World Countries: Agricultural Operations, 1982–1994

COUNTRY	CROPLAND				IRRIGATED LAND AS A % OF CROPLAND		ANNUAL FERTILIZER USE (kg/ha of cropland)		AGRICULTURAL EQUIPMENT			
									Tractors		Harvesters	
	Total Hectares (000) 1984	Hectares Per Capita (000) 1984	Total Hectares (000) 1994	Hectares Per Capita (000) 1994	1982–84	1992–94	1982–84	1992–94	Average Number 1992–94	Percent Change Since 1982–84	Average Number 1992–94	Percent Change Since 1982–84
WORLD	**1,208,584**	**0.25**	**1,238,812**	**0.22**	**17**	**19**	**107**	**113**	**25,951,850**	**11**	**4,134,798**	**8**
AFRICA	**179,281**	**0.34**	**190,022**	**0.27**	**6**	**6**	**19**	**18**	**539,098**	**15**	**40,322**	**(26)**
Algeria	7,510	0.35	8,043	0.29	4	7	27	15	95,562	78	9,786	92
Angola	3,400	0.44	3,500	0.33	2	2	4	3	10,300	0	X	X
Benin	1,818	0.47	1,880	0.36	0	0	4	11	140	25	X	X
Botswana	400	0.38	420	0.30	1	1	3	2	600	133	95	16
Burkina Faso	2,985	0.39	3,431	0.34	0	1	4	7	1,380	1,050	X	X
Burundi	1,180	0.26	1,110	0.19	1	1	2	3	170	55	2	100
Cameroon	6,965	0.72	7,040	0.55	0	0	6	4	500	(21)	X	X
Central African Republic	1,982	0.78	2,020	0.63	0	0	1	1	210	24	20	76
Chad	3,150	0.64	3,256	0.53	0	0	2	2	170	6	17	0
Congo (Zaire)	7,750	0.25	7,900	0.18	0	0	2	2	2,430	14	X	X
Congo Republic	158	0.08	170	0.07	1	1	18	11	707	4	73	88
Cote d'Ivoire	3,486	0.37	4,190	0.31	2	2	12	16	3,700	16	65	44
Egypt	2,493	0.05	3,265	0.05	100	100	341	264	72,648	61	2,370	11
Equatorial Guinea	230	0.78	230	0.59	0	0	X	0	100	2	X	X
Eritrea	X	X	519	0.17	X	5	X	0	567	X	27	X
Ethiopia	13,930	0.35	11,012	0.20	1	1	2	14	3,283	(16)	113	(24)
Gabon	452	0.58	460	0.44	1	1	6	1	1,500	15	X	X
Gambia	169	0.24	172	0.16	1	1	12	5	45	4	5	25
Ghana	3,900	0.31	4,500	0.27	0	0	2	3	4,100	12	540	50
Guinea	715	0.15	820	0.12	13	12	0	1	437	118	X	X
Guinea-Bissau	315	0.37	340	0.32	6	5	0	2	19	12	X	X
Kenya	4,285	0.22	4,520	0.17	1	1	18	31	14,000	73	650	50
Lesotho	297	0.20	320	0.16	1	1	15	19	1,850	22	35	17
Liberia	371	0.18	370	0.17	1	1	3	0	327	6	X	X
Libya	2,115	0.58	2,170	0.42	13	22	48	32	34,000	28	3,410	18
Madagascar	3,020	0.29	3,105	0.22	25	35	2	4	3,200	17	150	15
Malawi	1,450	0.21	1,700	0.18	1	2	31	16	1,420	11	X	X
Mali	2,053	0.27	3,000	0.29	3	3	24	8	2,383	99	50	9
Mauritania	195	0.11	208	0.09	25	24	3	19	333	7	X	X
Mauritius	107	0.11	106	0.10	15	17	253	275	370	10	X	X
Morocco	8,352	0.39	9,291	0.36	15	13	32	32	42,000	34	4,500	13
Mozambique	3,080	0.23	3,180	0.19	3	3	1	2	5,750	0	X	X
Namibia	662	0.58	750	0.50	1	1	0	0	3,150	17	X	X
Niger	3,535	0.55	4,500	0.51	1	2	1	1	180	39	X	X
Nigeria	30,873	0.38	32,700	0.30	1	1	9	9	11,900	25	X	X
Rwanda	1,108	0.19	1,150	0.22	0	0	2	2	90	7	X	X
Senegal	2,350	0.38	2,365	0.29	3	3	8	8	550	20	155	7
Sierra Leone	523	0.15	540	0.13	5	5	2	6	550	34	6	38

(Continued on next page)

COUNTRY	CROPLAND				IRRIGATED LAND AS A % OF CROPLAND		ANNUAL FERTILIZER USE (kg/ha of cropland)		AGRICULTURAL EQUIPMENT			
									Tractors		Harvesters	
	Total Hectares (000) 1984	Hectares Per Capita (000) 1984	Total Hectares (000) 1994	Hectares Per Capita (000) 1994	1982-84	1992-94	1982-84	1992-94	Average Number 1992-94	Percent Change Since 1982-84	Average Number 1992-94	Percent Change Since 1982-84
Somalia	1,020	0.13	1,020	0.11	16	19	4	0	1,980	7	X	X
South Africa	13,169	0.41	15,500	0.38	9	8	73	49	125,962	(26)	12,562	(64)
Sudan	12,608	0.60	12,975	0.50	15	15	3	4	10,500	6	1,600	70
Swaziland	179	0.28	191	0.23	39	35	47	70	4,060	17	X	X
Tanzania	2,895	0.14	3,675	0.13	4	4	12	10	6,600	(25)	X	X
Togo	2,360	0.80	2,400	0.60	0	0	3	5	370	42	X	X
Tunisia	4,990	0.70	4,813	0.55	6	8	17	18	30,158	15	3,070	19
Uganda	6,500	0.45	6,800	0.36	0	0	0	0	4,700	44	15	36
Zambia	5,158	0.82	5,273	0.67	0	1	11	11	6,000	19	293	8
Zimbabwe	2,808	0.35	2,878	0.26	3	4	52	59	20,833	31	720	11
NORTH AMERICA	**252,854**	**0.89**	**233,276**	**0.79**	**9**	**9**	**94**	**92**	**5,540,000**	**3**	**817,000**	**(2)**
Canada	46,055	1.80	45,500	1.56	1	2	52	50	740,000	8	155,000	(3)
United States	189,799	0.79	187,776	0.71	11	11	104	103	4,800,000	3	662,000	(1)
CENTRAL AMERICA	**38,107**	**0.39**	**41,112**	**0.34**	**17**	**20**	**65**	**58**	**293,717**	**14**	**32,612**	**13**
Belize	53	0.33	83	0.40	4	4	37	75	1,150	32	45	32
Costa Rica	518	0.20	530	0.16	17	23	186	243	7,000	15	1,190	12
Cuba	3,337	0.33	4,512	0.41	25	24	174	27	78,000	18	7,400	8
Dominican Republic	1,430	0.23	1,820	0.24	13	14	42	51	2,350	6	X	X
El Salvador	725	0.16	755	0.14	15	16	76	101	3,430	2	420	24
Guatemala	1,785	0.23	1,910	0.19	5	7	5	4	4,300	6	3,050	11
Haiti	902	0.16	910	0.13	8	9	4	6	230	21	X	X
Honduras	1,777	0.44	2,030	0.37	4	4	21	28	4,391	33	X	X
Jamaica	220	0.10	219	0.09	15	15	105	119	3,080	5	X	X
Mexico	24,688	0.33	24,730	0.28	20	25	67	62	172,000	14	19,500	18
Nicaragua	1,683	0.54	2,569	0.64	5	4	29	12	2,700	15	X	X
Panama	586	0.28	665	0.26	5	5	48	48	5,000	(6)	1,000	(26)
Trinidad and Tobago	118	0.10	122	0.10	18	19	52	49	2,650	6	X	X
SOUTH AMERICA	**104,544**	**0.40**	**114,901**	**0.37**	**7**	**8**	**44**	**60**	**1,214,092**	**21**	**121,732**	**16**
Argentina	27,200	0.91	27,200	0.79	6	6	5	17	280,000	38	50,000	10
Bolivia	2,217	0.38	2,370	0.33	6	4	2	4	5,350	19	123	1
Brazil	51,680	0.39	60,000	0.38	4	5	65	79	735,000	19	48,000	23
Chile	4,325	0.36	4,154	0.30	29	30	42	95	40,974	19	8,767	5
Colombia	5,264	0.18	6,220	0.18	8	14	69	92	27,000	(16)	2,850	27
Ecuador	2,505	0.28	3,036	0.27	16	9	29	51	8,900	20	780	24
Guyana	495	0.63	496	0.60	26	26	30	30	3,630	3	440	6
Paraguay	1,967	0.56	2,270	0.48	3	3	5	10	16,500	72	X	X
Peru	3,691	-0.19	4,140	0.18	33	43	21	41	12,933	8	X	X
Suriname	59	0.16	68	0.16	79	88	196	63	1,330	15	275	25

(Continued on next page)

COUNTRY	CROPLAND				IRRIGATED LAND AS A % OF CROPLAND		ANNUAL FERTILIZER USE (kg/ha of cropland)		AGRICULTURAL EQUIPMENT			
									Tractors		Harvesters	
	Total Hectares (000) 1984	Hectares Per Capita (000) 1984	Total Hectares (000) 1994	Hectares Per Capita (000) 1994	1982-84	1992-94	1982-84	1992-94	Average Number 1992-94	Percent Change Since 1982-84	Average Number 1992-94	Percent Change Since 1982-84
Uruguay	1,346	0.45	1,304	0.41	7	11	39	77	33,000	(2)	2,687	2
Venezuela	3,790	0.23	3,630	0.17	4	5	71	68	49,000	19	5,800	45
ASIA	X	X	621,590	0.18	X	X	X	X	7,197,097	X	1,856,053	X
Afghanistan	8,054	0.54	8,054	0.44	32	36	9	6	840	6	X	X
Armenia	X	X	600	0.17	X	49	X	12	15,852	X	2,085	X
Azerbaijan	X	X	2,000	0.27	X	54	X	20	31,000	X	4,233	X
Bangladesh	9,132	0.09	8,700	0.07	20	37	65	120	5,300	13	X	X
Bhutan	127	0.09	140	0.08	24	29	1	1	X	X	X	X
Cambodia	2,110	0.29	3,838	0.39	5	4	1	3	1,190	(3)	20	0
China	98,746	0.09	95,782	0.08	45	52	200	309	740,425	(12)	56,781	61
Georgia	X	X	1,127	0.21	X	44	X	28	20,567	X	1,271	X
India	169,078	0.22	169,700	0.19	24	29	49	60	1,196,268	136	3,167	21
Indonesia	25,934	0.16	30,171	0.16	17	15	72	80	46,976	353	237,148	1,460
Iran	15,540	0.33	18,500	0.28	40	39	60	54	229,335	107	5,488	79
Iraq	5,540	0.37	5,750	0.29	32	64	22	65	32,000	(5)	1,900	(24)
Israel	419	0.10	434	0.08	54	44	229	240	25,707	(5)	260	(25)
Japan	4,780	0.04	4,422	0.04	62	62	440	398	2,031,333	28	1,168,667	16
Jordan	351	0.10	405	0.08	12	17	43	35	7,378	55	75	17
Kazakhstan	X	X	34,978	2.08	X	6	X	3	208,586	X	79,628	X
Korea, North	1,945	0.10	2,000	0.09	63	73	398	377	75,000	22	X	X
Korea, South	2,153	0.05	2,033	0.05	61	65	361	472	76,555	903	66,373	1,030
Kuwait	3	X	5	0.00	57	73	150	200	100	329	X	X
Kyrgyzstan	X	X	1,420	0.32	X	74	X	20	26,059	X	3,698	X
Laos	810	0.23	900	0.19	16	16	0	2	890	24	X	X
Lebanon	298	0.11	306	0.10	29	29	170	114	3,800	27	97	7
Malaysia	5,300	0.35	7,604	0.39	6	5	105	159	34,617	283	X	X
Mongolia	1,336	0.72	1,320	0.55	4	6	13	1	8,107	(24)	2,001	(21)
Myanmar (Burma)	10,061	0.27	10,076	0.23	10	11	19	12	11,00	22	3,910	56
Nepal	2,329	0.14	2,743	0.13	29	35	18	33	4,600	71	X	X
Oman	47	0.03	63	0.03	92	95	20	159	150	44	45	246
Pakistan	20,330	0.21	21,510	0.16	76	80	62	98	289,433	119	1,633	173
Philippines	8,920	0.17	9,370	0.14	16	17	29	64	11,500	24	700	35
Saudi Arabia	2,423	0.20	3,800	0.21	34	34	108	95	8,633	220	2,233	244
Singapore	6	X	1	0.00	0	6	833	4,800	65	30	X	X
Sri Lanka	1,872	0.12	1,883	0.11	29	29	102	113	32,500	21	6	64
Syria	5,656	0.56	5,971	0.43	10	17	39	59	73,428	99	4,428	56
Tajikistan	X	X	860	0.15	X	85	X	81	31,909	X	1,100	X
Thailand	19,331	0.38	20,445	0.35	18	22	23	64	99,549	288	55,462	85
Turkey	27,413	0.56	27,771	0.46	11	15	56	54	745,248	44	11,408	44
Turkmenistan	X	X	1,480	0.37	X	90	X	84	55,015	X	14,672	(16)
United Arab Emirates	32	0.02	82	0.04	185	89	118	434	175	5	5	67
Uzbekistan	X	X	4,500	0.20	X	88	X	105	173,333	X	7,333	X
Vietnam	6,590	0.11	6,758	0.09	26	28	57	192	37,309	31	X	X

(Continued on next page)

COUNTRY	CROPLAND				IRRIGATED LAND AS A % OF CROPLAND		ANNUAL FERTILIZER USE (kg/ha of cropland)		AGRICULTURAL EQUIPMENT			
									Tractors		Harvesters	
	Total Hectares (000) 1984	Hectares Per Capita (000) 1984	Total Hectares (000) 1994	Hectares Per Capita (000) 1994	1982-84	1992-94	1982-84	1992-94	Average Number 1992-94	Percent Change Since 1982-84	Average Number 1992-94	Percent Change Since 1982-84
Yemen	1,465	0.16	1,540	0.11	20	29	13	7	5,943	12	55	22
EUROPE	**X**	**X**	**316,378**	**0.43**	**X**	**X**	**X**	**X**	**11,505,372**	**X**	**1,320,468**	**X**
Albania	713	0.25	702	0.21	55	49	132	21	9,050	(12)	902	(35)
Austria	1,522	0.20	1,513	0.19	0	0	257	168	345,753	6	24,992	(17)
Belarus	X	X	6,225	0.60	X	2	X	87	124,997	X	35,242	X
Belgium	775	0.08	794	0.08	0	0	538	402	112,047	(4)	9,190	(1)
Bosnia-Herzegovina	X	X	800	0.22	X	0	X	6	24,667	X	1,083	X
Bulgaria	4,135	0.46	4,219	0.49	29	24	227	52	37,145	(36)	6,802	(24)
Croatia	X	X	1,221	0.27	X	0	X	152	4,173	X	1,025	X
Czech Republic	X	X	3,386	0.33	X	1	X	100	43,151	X	6,823	X
Denmark	2,627	0.51	2,374	0.46	15	18	251	196	152,084	(14)	30,912	(17)
Estonia	X	X	1,144	0.76	X	0	X	36	47,742	X	2,300	X
Finland	2,439	0.50	2,593	0.51	2	2	208	148	232,000	(0)	38,000	(17)
France	19,145	0.35	19,488	0.34	5	8	302	242	1,366,667	(8)	154,000	4
Germany	12,428	0.16	12,037	0.15	4	4	381	241	1,307,000	(20)	136,927	(23)
Greece	3,952	0.40	3,502	0.34	25	37	167	152	230,409	36	6,173	(4)
Hungary	5,289	0.50	4,974	0.49	4	4	288	63	38,500	(31)	8,367	(32)
Iceland	7	0.02	6	0.02	0	0	3,806	3,433	10,595	(24)	2	(65)
Ireland	1,044	0.30	1,317	0.37	0	0	649	569	167,500	10	5,100	(2)
Italy	12,232	0.22	11,143	0.19	20	23	171	170	1,455,192	24	49,643	27
Latvia	X	X	1,740	0.68	X	0	X	55	58,200	X	35,233	X
Lithuania	X	X	3,017	0.81	X	0	X	26	58,853	X	8,000	X
Macedonia	X	X	661	0.31	X	11	X	88	47,976	X	1,251	X
Moldova	X	X	2,190	0.49	X	14	X	53	51,911	X	6,400	X
Netherlands	846	0.06	920	0.06	61	61	851	592	182,000	(0)	5,587	(4)
Norway	855	0.21	901	0.21	10	11	295	227	149,033	4	16,000	(8)
Poland	14,863	0.40	14,652	0.38	1	1	221	98	1,212,810	60	89,156	83
Portugal	3,153	0.32	3,000	0.31	20	21	69	83	152,530	41	4,362	(13)
Romania	10,754	0.47	9,926	0.43	24	31	142	48	155,380	(9)	42,533	(2)
Russian Fed.	X	X	132,302	0.89	X	4	X	11	1,227,187	X	344,923	X
Slovak Republic	X	X	1,611	0.30	X	19	X	59	21,220	X	3,919	X
Slovenia	X	X	286	0.15	X	1	X	286	84,000	X	1,300	X
Spain	20,512	0.54	20,129	0.51	15	18	80	95	777,104	31	48,946	12
Sweden	2,933	0.35	2,780	0.32	3	5	156	116	165,000	(12)	40,000	(17)
Switzerland	412	0.06	434	–0.06	6	6	437	336	114,000	10	4,000	(18)
Ukraine	X	X	34,357	0.66	X	8	X	33	445,282	X	99,218	X
United Kingdom	6,990	0.12	5,949	0.10	2	2	371	381	500,000	(5)	47,000	(16)
Yugoslavia (Serbia-Montenegro)	X	X	4,085	0.40	X	2	X	21	403,916	X	5,157	X
OCEANIA	**51,849**	**2.15**	**51,515**	**1.84**	**4**	**5**	**34**	**46**	**401,364**	**(5)**	**60,095**	**(3)**
Australia	47,239	3.07	47,205	2.67	4	5	26	35	315,000	(3)	56,500	(2)

(Continued on next page)

Sources: Food and Agriculture Organisation of the United Nations; United Nations Population Division; *World Resources 1998–99* (World Resources Institute).

COUNTRY	CROPLAND				IRRIGATED LAND AS A % OF CROPLAND		ANNUAL FERTILIZER USE (kg/ha of cropland)		AGRICULTURAL EQUIPMENT			
									Tractors		Harvesters	
	Total Hectares (000) 1984	Hectares Per Capita (000) 1984	Total Hectares (000) 1994	Hectares Per Capita (000) 1994	1982–84	1992–94	1982–84	1992–94	Average Number 1992–94	Percent Change Since 1982–84	Average Number 1992–94	Percent Change Since 1982–84
Fiji	190	0.28	260	0.34	1	1	83	69	7,017	43	X	X
New Zealand	3,500	1.09	3,071	0.87	7	9	147	212	76,000	(16)	3,200	(19)
Papua New Guinea	376	0.11	440	0.10	0	0	18	30	1,140	(8)	475	14
Solomon Islands	54	0.21	57	0.16	0	0	0	0	X	X	X	X
DEVELOPING COUNTRIES	763,202	0.21	799,479	0.18	21	23.	63	89	5,730,613	44	636,207	133
DEVELOPED COUNTRIES	677,068	0.61	667,273	0.57	9	10	121	78	20,221,237	4	3,498,591	(2)

NOTES: kg = kilogram; 1 kilogram = 2.205 lbs.
ha = hectare; 1 hectare = 2.471 acres

Table I

World Countries: Energy Production and Consumption

COUNTRY	COMMERCIAL ENERGY								TRADITIONAL ENERGY % of total energy use		CARBON DIOXIDE EMISSIONS				
	TOTAL PRODUCTION (1000 metric tons of oil equivalent)		TOTAL CONSUMPTION (1000 metric tons of oil equivalent)			PER CAPITA CONSUMPTION (Kilogram of oil equivalent)					Total Millions of Metric Tons		Per Capita Metric Tons (Kilogram of oil equivalent)		Kg of CO_2 per US$ of Gross Domestic Product
	1980	1995	1980	1995	Average Annual % Growth	1980	1995	Average Annual % Growth	1980	1995	1980	1995	1980	1995	1995
WORLD	6,273,572	8,385,643	6,325,980	8,244,516	3.2	1,456	1,474	3.3	7.1	6.8	13,585.7	22,700	3.4	4.1	1.2
AFRICA															
Algeria	66,730	109,257	12,078	24,346	4.2	647	866	1.4	12.3	8.6	4.8	1.8	1.9	0.6	1.0
Angola	7,700	26,189	937	959	0.5	133	89	(2.3)	47.2	59.5	5.3	4.6	0.8	0.4	1.4
Benin	X	232	149	107	(3.3)	43	20	(6.2)	84.9	92.5	0.5	0.6	0.1	0.1	0.3
Botswana	260	250	384	555	2.5	426	383	(0.8)	35.7	X	1.0	2.2	1.1	1.5	0.8
Burkina Faso	0	0	144	162	1.1	21	16	(1.5)	91.4	93.3	0.4	1.0	0.1	0.1	0.4
Burundi	1	5	58	144	6.4	14	23	3.5	92.7	88.8	0.1	0.2	0.0	0.0	0.2
Cameroon	2,855	5,380	774	1,556	3.3	89	117	0.4	69.4	77.3	3.9	4.1	0.4	0.3	0.4
Central African Republic	17	24	59	94	2.6	26	29	0.2	90.8	89.0	0.1	0.2	0.0	0.1	0.2
Chad	0	0	93	101	0.6	21	16	(1.8)	87.4	90.2	0.2	0.1	0.0	0.0	0.1
Congo (Zaire)	1,478	1,948	1,487	2,058	2.2	55	47	(1.1)	79.5	83.9	3.5	2.1	0.1	0.0	0.4
Congo Republic	3,387	9,031	262	367	2.6	157	139	(0.5)	55.9	61.0	0.4	1.3	0.2	0.5	0.5
Cote d'Ivoire	192	435	1,535	1,362	1.2	175	97	(2.4)	53.5	67.2	4.7	10.4	0.6	0.7	0.9
Egypt	33,374	59,287	15,178	34,678	5.4	371	596	2.9	5.0	3.3	45.2	91.7	1.1	1.6	1.6
Equatorial Guinea	X	X	X	X	X	X	X	X	X	X	X	X	X	X	X
Eritrea	X	X	X	X	X	X	X	X	X	X	X	X	X	X	X
Ethiopia	55	158	624	1,178	4.9	17	21	2.0	92.4	90.1	1.8	3.5	0.0	0.1	0.9
Gabon	9,090	18,703	831	644	(4.3)	1,203	587	(7.2)	33.6	51.8	4.8	3.5	6.9	3.2	0.7
Gambia	0	0	53	61	0.9	83	55	(2.9)	79.7	81.2	0.2	X	0.2	X	0.8
Ghana	554	526	1,303	1,564	2.7	121	92	(0.5)	68.2	79.0	2.4	4.0	0.2	0.2	0.6
Guinea	38	58	356	422	1.3	80	64	(1.4)	68.4	69.9	0.9	1.1	0.2	0.2	0.4
Guinea-Bissau	0	0	31	40	2.1	38	37	0.3	76.1	70.5	0.1	0.2	0.2	0.2	1.0
Kenya	91	518	1,991	2,907	3.5	120	109	0.3	75.4	76.1	6.2	6.7	0.4	0.3	0.7
Lesotho	X	X	X	X	X	X	X	X	X	X	X	X	X	X	X
Liberia	X	X	X	X	X	X	X	X	X	X	X	X	X	X	X
Libya	96,537	77,825	7,048	15,781	4.5	2,316	3,129	1.1	1.7	0.8	26.9	39.4	8.8	7.8	X
Madagascar	38	84	391	484	1.6	45	36	(1.2)	77.1	83.8	1.6	1.1	0.2	0.1	0.4
Malawi	99	154	334	374	1.6	54	38	(1.6)	89.1	86.8	0.7	0.7	0.1	0.1	0.5
Mali	21	42	164	207	1.7	25	21	(0.9)	85.2	87.4	0.4	0.5	0.1	0.0	0.2
Mauritania	0	0	214	231	0.5	138	102	(2.0)	0.7	X	0.6	3.1	0.4	1.3	2.7

(Continued on next page)

COUNTRY	COMMERCIAL ENERGY								TRADITIONAL ENERGY % of total energy use		CARBON DIOXIDE EMISSIONS				
	TOTAL PRODUCTION (1000 metric tons of oil equivalent)		TOTAL CONSUMPTION (1000 metric tons oil equivalent)			PER CAPITA CONSUMPTION (Kilogram of oil equivalent)					Total Millions of Metric Tons		Per Capita Metric Tons (Kilogram of oil equivalent)		Kg of CO_2 per US$ of Gross Domestic Product
	1980	1995	1980	1995	Average Annual % Growth	1980	1995	Average Annual % Growth	1980	1995	1980	1995	1980	1995	1995
Mauritius	21	34	339	435	2.8	351	388	1.7	44.1	41.6	0.6	1.5	0.6	1.3	0.5
Morocco	617	440	4,518	8,253	4.4	233	311	2.2	5.4	4.7	15.9	29.3	0.8	1.1	1.3
Mozambique	1,293	160	1,123	662	(1.6)	93	38	(3.5)	72.6	86.0	3.2	1.0	0.3	0.1	0.4
Namibia	X	X	X	X	X	X	X	X	X	X	X	X	X	X	X
Niger	14	56	210	330	2.0	38	38	(1.2)	78.0	79.6	0.6	1.1	0.1	0.1	0.5
Nigeria	105,512	104,475	9,879	18,393	3.4	139	165	0.4	63.7	56.6	68.1	90.7	1.0	0.8	2.7
Rwanda	29	46	190	211	(0.7)	37	33	(2.6)	84.8	85.7	0.3	0.5	0.1	0.1	0.4
Senegal	0	46	875	866	(0.3)	158	104	(3.0)	48.6	55.9	2.8	3.1	0.5	0.4	0.6
Sierra Leone	0	0	310	326	0.5	96	72	(1.7)	63.5	69.4	0.6	0.4	0.2	0.1	0.6
Somalia	X	X	X	X	X	X	X	X	X	X	X	X	X	X	X
South Africa	66,740	116,160	59,051	88,882	1.8	2,175	2,405	(0.2)	4.5	3.9	211.3	305.8	7.8	8.3	3.4
Sudan	58	81	1,140	1,745	3.3	61	65	0.9	76.4	76.4	3.3	3.5	0.2	0.1	0.2
Swaziland	X	X	X	X	X	X	X	X	X	X	X	X	X	X	X
Tanzania	86	135	1,023	947	0.8	55	32	(2.3)	83.7	89.6	1.9	2.4	0.1	0.1	0.6
Togo	1	0	195	185	0.9	75	45	(2.1)	38.3	73.1	0.6	0.7	0.2	0.2	1.2
Tunisia	6,149	4,579	3,083	5,314	4.0	483	591	1.7	15.4	12.9	9.4	15.3	1.5	1.7	1.2
Uganda	153	185	320	430	2.8	25	22	0.0	87.2	89.2	0.6	1.0	0.1	0.1	0.1
Zambia	1,146	898	1,685	1,302	(2.1)	294	145	(5.0)	54.6	71.2	3.5	2.4	0.6	0.3	1.1
Zimbabwe	2,024	3,567	2,797	4,673	4.4	399	424	1.3	33.6	27.4	9.6	9.7	1.4	0.9	1.5
NORTH AMERICA															
Canada	207,359	350,629	192,942	233,328	1.6	7,845	7,879	0.3	0.6	0.6	420.9	435.7	17.1	14.7	0.9
United States	1,546,307	1,655,644	1,801,406	2,078,265	1.3	7,928	7,905	0.3	1.2	4.2	4,515.3	5,468.6	19.9	20.8	1.0
CENTRAL AMERICA															
Belize	X	X	X	X	X	X	X	X	X	X	X	X	X	X	X
Costa Rica	181	380	949	1,971	6.0	415	584	3.3	40.4	12.7	2.5	5.2	1.1	1.6	0.6
Cuba	293	1,223	9,645	10,437	0.1	992	949	(0.9)	28.1	19.7	30.7	29.1	3.2	2.6	X
Dominican Republic	50	171	2,211	3,801	4.3	388	486	2.1	28.3	12.1	6.4	22.8	1.1	1.5	1.6
El Salvador	407	703	1,004	2,332	5.7	221	410	4.2	50.3	42.9	2.1	5.2	0.5	0.9	0.9
Guatemala	230	589	1,443	2,191	3.6	209	206	0.6	53.1	59.9	4.5	7.2	0.6	0.7	0.7
Haiti	19	32	241	357	0.1	45	50	(1.8)	82.4	80.3	0.8	0.6	0.1	0.1	0.5
Honduras	67	235	636	1,401	5.1	174	236	1.9	61.2	49.3	2.1	3.9	0.6	0.7	0.7
Jamaica	10	10	2,164	3,003	2.7	1,015	1,191	1.6	6.2	8.0	8.4	9.1	4.0	3.6	2.4
Mexico	149,365	201,957	98,904	133,371	2.2	1,486	1,456	0.0	4.4	4.4	255.0	357.8	3.8	3.9	2.1
Nicaragua	44	302	696	1,159	3.4	248	265	0.3	50.4	45.8	2.0	2.7	0.7	0.6	0.7
Panama	83	202	1,419	1,783	1.6	725	678	(0.4)	26.4	19.4	3.5	6.9	1.8	2.6	1.0
Trinidad and Tobago	13,127	12,991	3,860	6,925	4.0	3,567	5,381	2.8	1.8	1.0	16.7	17.1	15.4	13.3	3.5

(Continued on next page)

COUNTRY	COMMERCIAL ENERGY								TRADITIONAL ENERGY % of total energy use		CARBON DIOXIDE EMISSIONS				
	TOTAL PRODUCTION (1000 metric tons of oil equivalent)		TOTAL CONSUMPTION (1000 metric tons of oil equivalent)			PER CAPITA CONSUMPTION (Kilogram of oil equivalent)					Total Millions of Metric Tons		Per Capita Metric Tons (Kilogram of oil equivalent)		Kg of CO$_2$ per US$ of Gross Domestic Product
	1980	1995	1980	1995	Average Annual % Growth	1980	1995	Average Annual % Growth	1980	1995	1980	1995	1980	1995	1995
SOUTH AMERICA															
Argentina	36,661	66,055	39,716	53,016	1.9	1,413	1,525	0.5	6.5	4.0	107.5	129.5	3.8	3.7	1.0
Bolivia	3,553	4,478	1,599	2,939	3.2	299	396	1.0	19.8	12.8	4.5	10.5	0.8	1.4	1.8
Brazil	25,777	73,172	73,041	122,928	4.2	602	772	2.3	41.2	27.5	183.4	249.2	1.5	1.6	0.8
Chile	3,871	4,361	7,732	15,131	5.4	694	1,065	3.6	14.5	13.3	27.9	44.1	2.5	3.1	1.2
Colombia	13,047	54,361	13,962	24,120	3.5	501	655	1.6	21.4	21.1	39.8	67.5	1.4	1.8	1.3
Ecuador	10,774	20,967	4,209	6,343	2.6	529	553	0.1	26.5	14.8	13.4	22.6	1.7	2.0	1.6
Guyana	X	X	X	X	X	X	X	X	X	X	X	X	X	X	X
Paraguay	58	3,578	544	1,487	7.1	173	308	4.1	66.1	51.5	1.5	3.8	0.5	0.8	0.7
Peru	11,188	8,388	8,233	10,035	0.6	476	421	(1.5)	18.7	22.9	23.5	30.6	1.4	1.3	1.3
Suriname	X	X	X	X	X	X	X	X	X	X	X	X	X	X	X
Uruguay	233	477	2,206	2,035	0.7	757	639	0.1	20.4	26.7	5.8	5.4	2.0	1.7	0.6
Venezuela	132,919	187,498	35,011	47,140	1.7	2,354	2,158	(0.9)	1.0	1.2	89.6	180.2	6.0	8.3	3.1
ASIA															
Afghanistan	X	X	X	X	X	X	X	X	X	X	X	X	X	X	X
Armenia	X	244	1,070	1,671	(1.8)	346	444	(3.1)	X	X	X	3.6	X	1.0	3.4
Azerbaijan	14,821	14,719	15,001	13,033	(3.9)	2,433	1,735	(5.1)	X	X	X	42.6	X	5.7	14.6
Bangladesh	1,113	5,962	2,809	8,061	7.4	32	67	5.1	67.8	49.9	7.6	20.9	0.1	0.2	0.9
Bhutan	X	X	X	X	X	X	X	X	X	X	X	X	X	X	X
Cambodia	13	22	393	517	2.1	60	52	(1.0)	71.2	75.3	0.3	0.5	0.0	0.0	0.4
China	428,693	866,556	413,176	860,520	5.1	421	707	3.7	8.0	5.6	1,476.8	3,192.5	1.5	2.7	5.5
Georgia	4,706	478	4,474	1,850	(3.3)	882	342	(3.7)	X	1.3	X	7.7	X	1.4	X
India	73,760	196,941	93,897	241,291	6.5	137	260	4.4	34.7	23.3	347.3	908.7	0.5	1.0	2.2
Indonesia	94,717	169,325	25,904	85,785	8.9	175	442	7.0	51.6	29.9	94.6	296.1	0.6	1.5	2.1
Iran	83,430	216,406	38,347	84,069	6.3	980	1,374	3.2	1.6	0.8	116.1	263.8	3.0	4.3	1.4
Iraq	136,616	31,100	12,003	25,061	4.1	923	1,206	0.8	0.2	0.1	44.0	99.0	3.4	4.8	X
Israel	151	562	8,607	16,650	5.0	2,219	3,003	2.6	X	X	21.1	46.3	5.4	8.4	0.8
Japan	43,247	99,468	346,567	497,231	2.8	2,968	3,964	2.3	0.1	0.5	907.4	1,126.8	7.8	9.0	0.4
Jordan	0	192	1,713	4,323	5.2	785	1,031	0.7	X	X	4.7	13.3	2.2	3.2	1.7
Kazakhstan	76,799	64,354	75,799	55,432	(3.1)	5,153	3,337	(3.8)	X	0.1	X	221.5	X	13.3	13.8
Korea, North	28,275	21,538	30,932	24,600	(1.2)	1,751	1,113	(2.6)	2.7	3.9	124.9	257.0	7.1	11.6	X
Korea, South	9,644	20,570	41,426	145,099	9.6	1,087	3,225	8.4	5.7	0.7	125.2	373.6	3.3	8.3	1.5
Kuwait	94,084	111,227	9,561	14,494	0.3	6,953	9,381	0.2	X	X	24.7	48.7	18.0	31.5	1.7
Kyrgyzstan	2,190	1,377	1,938	2,315	5.0	534	513	3.4	X	X	X	5.5	X	1.2	4.9
Lao PDR	236	220	107	184	2.6	33	40	0.1	86.6	85.1	0.2	0.3	0.1	0.1	0.2
Lebanon	73	69	2,376	4,486	3.2	791	1,120	1.2	4.3	2.6	6.2	13.3	2.1	3.3	2.4
Malaysia	15,049	62,385	9,522	33,252	9.8	692	1,655	7.0	14.4	6.6	28.0	106.6	2.0	5.3	1.7
Mongolia	1,195	2,190	1,943	2,576	1.8	1,168	1,045	(0.9)	14.0	3.6	6.8	8.5	4.1	3.4	X

(Continued on next page)

COUNTRY	COMMERCIAL ENERGY								TRADITIONAL ENERGY % of total energy use		CARBON DIOXIDE EMISSIONS				
	TOTAL PRODUCTION (1000 metric tons of oil equivalent)		TOTAL CONSUMPTION (1000 metric tons of oil equivalent)			PER CAPITA CONSUMPTION (Kilogram of oil equivalent)					Total Millions of Metric Tons		Per Capita Metric Tons (Kilogram of oil equivalent)		Kg of CO₂ per US$ of Gross Domestic Product
	1980	1995	1980	1995	Average Annual % Growth	1980	1995	Average Annual % Growth	1980	1995	1980	1995	1980	1995	1995
Myanmar (Burma)	1,940	2,167	1,858	2,234	0.2	55	50	(1.7)	66.5	69.4	4.8	7.0	0.1	0.2	X
Nepal	15	97	174	700	9.3	12	33	6.5	94.8	88.9	0.5	1.5	0.0	0.1	0.3
Oman	14,756	45,403	1,010	4,013	9.2	917	1,880	4.6	X	X	5.9	11.4	5.3	5.3	0.9
Pakistan	6,970	18,612	11,451	31,536	7.0	139	243	3.8	27.2	20.2	31.6	85.4	0.4	0.7	1.7
Philippines	2,789	6,006	13,357	21,542	3.6	276	307	0.9	35.8	30.5	36.5	61.2	0.8	0.9	1.4
Saudi Arabia	533,071	469,820	35,355	82,742	5.2	3,772	4,360	0.3	X	X	130.7	254.3	14.0	13.4	2.6
Singapore	0	0	6,049	21,389	10.0	2,651	7,162	8.1	0.0	0.0	30.1	63.7	13.2	21.3	1.5
Sri Lanka	127	383	1,411	2,469	2.7	96	136	1.3	54.3	48.4	3.4	5.9	0.2	0.3	0.6
Syria	9,495	34,287	5,343	14,121	5.9	614	1,001	2.5	0.1	0.0	19.3	46.0	2.2	3.3	2.6
Tajikistan	1,986	1,325	1,650	3,283	8.9	416	563	6.1	X	X	X	3.7	X	0.6	2.5
Thailand	535	19,430	12,093	52,125	11.1	259	878	9.4	48.3	32.7	40.1	175.0	0.9	2.9	1.6
Turkey	17,190	26,079	31,314	62,187	4.9	704	1,009	2.6	18.0	3.1	76.3	165.9	1.7	2.7	1.5
Turkmenistan	8,034	32,589	7,948	13,737	X	2,778	3,047	(9.8)	X	X	X	28.3	X	6.3	X
United Arab Emirates	93,915	138,821	8,576	28,454	7.5	8,222	11,567	1.6	X	X	36.3	68.3	34.8	27.8	X
Uzbekistan	4,615	49,135	4,821	46,543	11.6	302	2,043	8.9	X	X	X	98.9	X	4.3	7.3
Vietnam	2,728	13,808	4,024	7,694	4.1	75	104	1.8	53.5	49.1	16.8	31.7	0.3	0.4	0.5
Yemen	X	17,394	1,364	2,933	5.3	160	192	1.2	X	X	1.2	X	0.1	X	X
EUROPE															
Albania	3,053	940	2,674	1,020	(6.4)	1,001	324	(7.7)	12.3	8.6	4.8	1.8	1.8	0.6	1.0
Austria	7,654	8,481	23,449	26,383	1.3	3,105	3,279	0.9	1.4	2.8	52.2	59.3	6.9	7.4	0.4
Belarus	2,566	2,793	2,385	23,808	10.3	247	2,305	9.7	X	0.8	X	59.3	X	5.7	3.4
Belgium	7,986	11,628	46,100	52,378	1.6	4,682	5,167	1.3	0.2	0.9	127.2	103.8	12.9	10.2	0.6
Bosnia-Herzegovina	X	470	X	1,595	X	X	364	X	X	9.9	X	1.8	X	0.4	X
Bulgaria	7,541	9,810	28,476	22,878	(2.5)	3,213	2,724	(2.1)	0.7	0.8	75.3	56.7	8.5	6.7	2.5
Croatia	X	3,917	X	6,852	X	X	1,435	X	X	3.0	X	17.0	X	3.6	X
Czech Republic	40,002	20,448	45,766	39,013	(1.2)	4,473	3,776	(1.2)	X	0.5	X	112.0	X	10.8	3.4
Denmark	896	15,497	19,734	20,481	0.7	3,852	3,918	0.6	0.3	3.3	62.9	54.9	12.3	10.5	0.5
Estonia	X	3,117	X	5,126	X	X	3,454	X	X	2.3	X	16.4	X	11.1	4.3
Finland	6,912	12,911	25,002	28,670	1.5	5,235	5,613	1.1	3.8	5.1	54.9	51.0	11.5	10.0	0.5
France	46,829	126,868	190,109	241,322	2.1	3,528	4,150	1.6	1.3	1.0	482.7	340.1	9.0	5.8	0.3
Germany	184,238	142,712	358,995	339,287	(0.2)	4,585	4,156	(0.5)	X	0.7	X	835.1	X	10.2	X
Greece	3,696	9,053	15,960	23,698	3.2	1,655	2,266	2.7	2.8	1.5	51.7	76.3	5.4	7.3	1.5
Hungary	14,442	13,295	28,556	25,103	(1.0)	2,667	2,454	(0.7)	2.1	1.8	82.5	55.9	7.7	5.5	2.3
Iceland	X	X	X	X	X	X	X	X	X	X	X	X	X	X	X
Ireland	1,894	3,601	8,484	11,461	2.2	2,495	3,196	2.0	0.1	0.2	25.2	32.2	7.4	9.0	0.7
Italy	19,644	28,645	138,629	161,360	1.4	2,456	2,821	1.3	0.7	1.9	371.9	410.0	6.6	7.2	0.5

(Continued on next page)

COUNTRY	COMMERCIAL ENERGY								TRADITIONAL ENERGY % of total energy use		CARBON DIOXIDE EMISSIONS				
	TOTAL PRODUCTION (1000 metric tons of oil equivalent)		TOTAL CONSUMPTION (1000 metric tons of oil equivalent)			PER CAPITA CONSUMPTION (Kilogram of oil equivalent)					Total Millions of Metric Tons		Per Capita Metric Tons (Kilogram of oil equivalent)		Kg of CO_2 per US$ of Gross Domestic Product
	1980	1995	1980	1995	Average Annual % Growth	1980	1995	Average Annual % Growth	1980	1995	1980	1995	1980	1995	1995
Latvia	261	322	566	3,702	22.9	222	1,471	22.9	X	18.0	X	9.3	X	3.7	1.9
Lithuania	186	3,316	11,353	8,510	(3.2)	3,326	2,291	(3.8)	X	5.6	X	14.8	X	4.0	2.1
Macedonia	X	1,621	X	2,572	X	X	1,308	X	X	6.9	X	X	X	X	X
Moldova	35	24	X	4,177	X	X	963	X	X	0.5	X	10.8	X	2.5	X
Netherlands	71,830	65,705	65,000	73,292	1.4	4,594	4,741	0.8	0.0	0.5	152.6	135.9	10.8	8.8	0.5
Norway	55,743	182,428	18,819	23,715	1.8	4,600	5,439	1.4	0.8	1.1	90.4	72.5	22.1	16.6	0.7
Poland	120,774	94,666	124,557	94,472	(2.0)	3,501	2,448	(2.5)	0.4	1.1	456.2	338.0	12.8	8.8	5.1
Portugal	1,481	1,870	10,291	19,245	4.6	1,054	1,939	4.6	1.1	0.7	27.1	51.9	2.8	5.2	1.0
Romania	51,631	30,008	63,751	44,026	(2.9)	2,872	1,941	(3.1)	1.5	21.5	191.8	121.1	8.6	5.3	3.9
Russian Federation	749,289	928,870	764,349	604,461	(3.0)	5,499	4,079	(3.4)	X	1.1	X	1,818.0	X	12.3	6.1
Slovak Republic	3,251	4,846	20,648	17,447	(1.3)	4,142	3,272	(1.7)	X	0.5	X	38.0	X	7.1	2.3
Slovenia	1,623	2,578	4,269	5,583	0.7	2,245	2,806	0.4	X	0.8	X	11.7	X	5.9	X
Spain	15,781	31,422	68,583	103,491	3.2	1,834	2,639	2.9	0.5	0.6	200.0	231.6	5.3	5.9	0.6
Sweden	16,133	31,549	40,984	50,658	1.3	4,932	5,736	0.9	3.9	2.5	71.4	44.6	8.6	5.0	0.3
Switzerland	7,030	10,961	20,814	25,142	1.7	3,294	3,571	0.9	1.1	2.1	40.9	38.9	6.5	5.5	0.2
Ukraine	109,708	80,700	97,893	161,586	2.1	1,956	3,136	1.9	X	0.4	X	438.2	X	8.5	X
United Kingdom	197,738	254,967	102,268	221,911	1.0	3,571	3,786	0.7	0.0	1.1	585.1	542.1	10.4	9.3	0.7
Yugoslavia (Serbia-Montenegro)	X	11,295	X	11,865	X	X	1,125	X	X	X	X	33.0	X	3.1	X
OCEANIA															
Australia	86,096	186,625	70,372	94,200	2.2	4,790	5,215	0.7	2.1	3.8	202.8	289.8	13.8	16.0	1.1
Fiji	X	X	X	X	X	X	X	X	X	X	X	X	X	X	X
New Zealand	5,592	12,436	9,190	15,409	3.9	2,952	4,290	3.1	0.2	X	17.6	27.4	5.6	7.6	0.6
Papua New Guinea	80	2,500	705	1,000	2.4	228	232	0.2	64.1	58.9	1.8	2.5	0.6	0.6	0.5
Solomon Islands	X	X	X	X	X	X	X	X	X	X	X	X	X	X	X
DEVELOPING COUNTRIES	3,488,398	4,811,025	2,522,195	3,569,396	5.9	1,665	1,415	2.3	34.3	31.7	4,813.6	11,577.5	2.7	3.1	2.6
DEVELOPED COUNTRIES	2,785,174	3,574,618	3,803,785	4,675,120	1.7	4,611	5,123	1.1	1.1	2.5	8,772.1	11,122.7	12.0	12.5	0.7

Sources: United Nations Statistical Division; *World Development Indicators* (World Bank, 1998); *World Resources 1998–1999* (World Resources Institute, 1998).

Table J
World Countries: Land Use and Change

COUNTRY	Domestic and Forested Land				Land Use (in Thousands of Hectares)							
	Domestic Land as a % of Land Area	Forests as a % of Original Forest		Percent Frontier Forests Threatened	Cropland		Permanent Pasture		Forest and Woodland		Other Land	
	1994	Current Forests	Frontier Forests	1994	1992–94	% Change Since 1982–84	1992–94	% Change Since 1982–84	1992–94	% Change Since 1982–84	1992–94	% Change Since 1982–84
WORLD	37	53.4	2.17	39.5	1,465,814	2.0	3,419,203	3.2	4,177,088	(2.2)	3,992,533	(1.0)
AFRICA	**36**	**33.9**	**7.8**	**76.8**	**189,803**	**6.5**	**889,350**	**0.0**	**713,405**	**(0.3)**	**1,171,024**	**(0.8)**
Algeria	17	12.0	0.0	0.0	8,088	9.1	32,014	(2.8)	3,949	(10.0)	195,197	0.4
Angola	46	15.3	0.0	0.0	3,500	2.9	54,000	0.0	23,000	(0.9)	44,170	0.2
Benin	21	3.5	0.0	0.0	1,880	3.9	42	0.0	3,400	(11.0)	5,340	7.0
Botswana	46	100.0	0.0	0.0	420	5.0	25,600	0.0	26,500	0.0	4,153	(0.5)
Burkina Faso	34	0.0	0.0	0.0	3,465	18.0	6,000	0.0	13,800	0.0	4,082	(11.7)
Burundi	86	3.5	0.0	0.0	1,120	(5.1)	1,080	9.1	325	0.0	43	(41.1)
Cameroon	19	42.4	7.9	97.4	7,040	1.2	2,000	0.0	35,900	0.0	1,600	(4.9)
Central African Republic	8	15.9	4.4	100.0	2,020	2.5	3,000	0.0	46,700	0.0	10,578	(0.5)
Chad	38	0.0	0.0	0.0	3,256	3.4	45,000	0.0	32,400	0.0	45,264	(0.2)
Congo (Zaire)	30	60.4	15.6	70.4	170	9.9	10,000	0.0	19,900	0.0	4,080	(0.4)
Congo Republic	10	67.8	28.7	64.6	7,900	2.5	15,000	0.0	166,000	0.0	37,805	(0.5)
Cote d'Ivoire	54	9.9	2.2	100.0	4,031	22.5	13,000	0.0	9,600	(5.9)	5,149	(3.0)
Egypt	3	0.0	0.0	0.0	3,137	26.5	X	X	34	9.7	96,374	(0.7)
Equatorial Guinea	12	38.4	0.0	0.0	230	0.0	104	0.0	1,830	0.0	641	0.0
Eritrea	75	X	X	X	366	X	4,622	X	523	X	1,155	X
Ethiopia	31	17.3	0.0	0.0	12,197	X	28,267	X	13,633	X	49,269	X
Gabon	20	90.4	32.4	0.0	460	1.8	4,700	0.0	19,900	(0.4)	707	11.3
Gambia	37	61.9	0.0	100.0	165	(10.6)	194	2.1	94	(6.0)	547	4.1
Ghana	57	8.6	0.0	0.0	4,407	15.0	8,400	0.0	9,300	(3.1)	647	(29.7)
Guinea	47	5.0	0.0	0.0	787	10.2	10,700	0.0	6,700	0.0	6,385	(1.1)
Guinea-Bissau	50	33.7	0.0	0.0	340	10.3	1,080	0.0	1,070	0.0	322	(9.0)
Kenya	45	18.5	0.0	0.0	4,520	5.6	21,300	0.0	16,800	0.0	14,294	(1.6)
Lesotho	76	0.0	0.0	0.0	320	10.6	2,000	0.0	X	X	715	(4.1)
Liberia	25	44.2	0.0	0.0	371	(0.1)	2,000	0.0	4,600	0.0	2,661	0.0
Libya	9	0.0	0.0	0.0	2,170	3.1	13,300	0.3	840	33.0	159,644	(0.2)
Madagascar	47	13.1	0.0	0.0	3,105	3.1	24,000	0.0	23,200	0.0	7,849	(1.2)
Malawi	38	0.0	0.0	0.0	1,700	19.9	1,840	0.0	3,700	(1.1)	2,168	(10.0)
Mali	27	0.0	0.0	0.0	2,569	25.1	30,000	0.0	11,800	(1.7)	77,650	(0.4)
Mauritania	38	0.0	0.0	0.0	208	6.7	39,250	0.0	4,410	(2.0)	58,654	0.1
Mauritius	56	X	X	X	106	(0.9)	7	0.0	44	(24.1)	46	48.4
Morocco	68	7.3	0.0	0.0	9,686	13.9	20,933	0.2	8,613	9.9	5,397	(28.8)
Mozambique	60	13.6	0.0	0.0	3,180	3.2	44,000	0.0	17,300	0.0	13,929	(0.7)
Namibia	47	95.3	0.0	0.0	704	6.6	38,000	0.0	12,500	0.0	31,125	(0.1)
Niger	12	0.0	0.0	0.0	4,035	13.9	10,400	13.1	2,500	0.0	109,695	(1.5)
Nigeria	80	10.7	0.6	100.0	32,579	6.1	40,000	0.0	14,300	0.0	4,198	(3.9)
Rwanda	75	16.1	0.0	0.0	1,150	5.4	695	(0.7)	250	(10.6)	372	(12.7)
Senegal	42	16.0	0.0	0.0	2,355	0.2	5,700	0.0	7,467	0.0	3,731	3.6
Sierra Leone	38	9.7	0.0	0.0	540	4.4	2,201	(0.1)	1,947	(1.8)	2,474	(2.6)
Somalia	70	0.0	0.0	0.0	1,026	1.1	43,000	0.0	16,000	2.5	2,708	(27.2)

(Continued on next page)

| | Domestic and Forested Land | | | | Land Use (in Thousands of Hectares) | | | | | | | |
| COUNTRY | Domestic Land as a % of Land Area | Forests as a % of Original Forest | | Percent Frontier Forests Threatened | Cropland | | Permanent Pasture | | Forest and Woodland | | Other Land | |
	1994	Current Forests	Frontier Forests	1994	1992–94	% Change Since 1982–84	1992–94	% Change Since 1982–84	1992–94	% Change Since 1982–84	1992–94	% Change Since 1982–84
South Africa	79	0.2	0.0	0.0	15,200	15.4	81,433	0.1	8,200	0.0	17,271	(10.8)
Sudan	52	0.0	0.0	0.0	12,975	3.3	110,000	12.2	42,367	(1.5)	72,258	(14.0)
Swaziland	73	0.0	0.0	0.0	191	24.3	1,070	(5.1)	119	16.6	340	0.9
Tanzania	44	9.1	0.0	0.0	3,660	23.7	35,000	0.0	33,067	(1.9)	16,632	(0.4)
Togo	48	7.0	0.0	0.0	2,420	2.5	200	0.0	900	(11.8)	1,919	3.2
Tunisia	51	4.7	0.0	0.0	4,882	(0.1)	3,416	1.8	666	17.1	6,602	(1.8)
Uganda	43	4.4	0.0	0.0	6,780	9.1	1,800	0.0	6,300	5.0	5,085	(14.6)
Zambia	47	70.1	0.0	0.0	5,273	2.2	30,000	0.0	32,000	6.7	7,066	(23.0)
Zimbabwe	52	67.3	0.0	0.0	2,876	2.5	17,190	0.5	8,800	(7.4)	9,819	5.8
NORTH AMERICA	**27**	**28.2**	**5.3**	**63.1**	**233,276**	**(1.1)**	**267,072**	**(1.2)**	**749,290**	**2.9**	**588,371**	**(2.6)**
Canada	8	91.2	56.5	20.9	45,500	(1.3)	27,900	(3.1)	453,300	3.9	395,397	(3.8)
United States	47	60.2	6.3	84.7	187,776	(1.1)	239,172	(1.0)	295,990	1.5	192,974	(0.0)
CENTRAL AMERICA	**53**	**54.5**	**9.7**	**87.0**	**40,053**	**5.4**	**98,503**	**6.2**	**74,524**	**1.2**	**85,910**	**(9.2)**
Belize	6	95.7	35.5	66.1	81	52.2	49	11.4	2,100	0.0	50	(39.4)
Costa Rica	56	34.9	9.5	100.0	530	2.9	2,340	6.9	1,570	(3.4)	666	(14.4)
Cuba	61	28.8	0.0	0.0	3,745	12.4	2,705	0.8	2,505	(6.2)	2,027	(11.8)
Dominican Republic	81	25.1	0.0	0.0	1,743	21.9	2,090	(0.1)	603	(4.1)	401	(41.6)
El Salvador	65	9.9	0.0	0.0	742	2.3	600	(1.6)	105	(13.9)	625	1.7
Guatemala	42	46.2	2.2	100.0	1,858	4.1	2,568	65.5	5,212	15.8	1,171	(60.9)
Haiti	51	0.8	0.0	0.0	910	1.2	495	(2.2)	140	0.0	1,211	0.0
Honduras	32	51.6	16.0	100.0	1,967	11.0	1,524	1.6	6,000	0.0	1,698	(11.4)
Jamaica	44	35.6	0.0	0.0	219	(1.9)	257	0.0	185	(3.8)	422	2.8
Mexico	55	63.4	8.1	77.0	24,370	0.2	79,000	6.0	48,700	4.3	38,439	(14.6)
Nicaragua	61	44.3	21.6	100.0	2,480	55.6	4,815	0.0	3,200	(22.9)	1,645	4.0
Panama	29	62.0	34.8	100.0	662	14.0	1,483	9.1	3,260	(16.0)	2,038	25.6
Trinidad and Tobago	26	35.5	0.0	0.0	122	3.4	11	0.0	235	3.5	145	(7.6)
SOUTH AMERICA	**35**	**69.1**	**45.6**	**54.0**	**113,116**	**9.0**	**495,341**	**3.0**	**934,860**	**0.6**	**209,471**	**(12.3)**
Argentina	62	59.5	6.3	99.9	27,200	0.0	142,000	(0.6)	50,900	0.0	53,569	1.7
Bolivia	27	77.2	43.6	96.9	2,370	8.9	26,500	(1.7)	58,000	0.0	21,565	1.2
Brazil	29	66.4	42.2	47.8	58,667	15.0	185,600	5.9	557,667	0.8	43,718	(34.1)
Chile	23	40.6	54.5	76.0	4,216	(2.0)	13,100	(0.4)	16,500	0.0	41,030	0.3
Colombia	45	53.5	36.4	18.7	6,073	15.7	40,083	(0.0)	53,167	(3.3)	4,547	29.1
Ecuador	29	66.4	36.9	99.5	3,010	20.8	5,009	11.1	15,600	0.6	4,065	(21.6)
Guyana	9	97.4	81.8	41.1	496	0.2	1,230	0.3	16,456	0.5	1,503	(5.7)
Paraguay	60	44.5	0.0	0.0	2,270	17.6	21,700	27.6	12,850	(31.6)	2,910	43.6
Peru	24	86.6	56.7	99.6	3,767	3.3	27,120	0.0	84,800	(0.1)	12,213	(1.4)
Suriname	1	95.6	92.2	21.7	68	19.3	21	12.5	15,000	0.8	511	(21.3)
Uruguay	85	0.0	0.0	0.0	1,304	(5.0)	13,520	(0.5)	930	0.0	1,727	8.8
Venezuela	25	83.6	59.3	37.3	3,663	(3.0)	18,742	2.7	45,000	22.3	21,300	(28.7)
ASIA	**51**	**28.2**	**5.3**	**63.1**	**520,175**	**X**	**1,051,311**	**X**	**556,996**	**X**	**956,913**	**X**
Afghanistan	58	6.5	0.0	0.0	8,054	0.0	30,000	X	1,700	(10.5)	25,455	0.8
Armenia	46	21.1	0.0	0.0	582	X	688	X	413	X	1,137	X
Azerbaijan	48	32.0	0.0	0.0	1,967	X	2,200	X	950	X	3,543	X

(Continued on next page)

| | Domestic and Forested Land | | | | Land Use (in Thousands of Hectares) | | | | | | | |
| | Domestic Land as a % of Land Area | Current Forests | Forests as a % of Original Forest — Frontier Forests | Percent Frontier Forests Threatened | Cropland | | Permanent Pasture | | Forest and Woodland | | Other Land | |
COUNTRY	1994	1994	1994	1994	1992–94	% Change Since 1982–84	1992–94	% Change Since 1982–84	1992–94	% Change Since 1982–84	1992–94	% Change Since 1982–84
Bangladesh	71	7.9	3.8	100.0	8,849	(3.1)	600	0.0	1,891	(11.3)	1,677	45.2
Bhutan	9	61.8	24.0	100.0	136	7.7	273	2.5	3,100	14.2	1,191	(25.2)
Cambodia	30	65.1	10.3	100.0	3,832	81.9	1,500	158.5	12,200	(7.3)	120	93.4
China	53	21.6	1.8	92.8	95,145	(3.6)	400,000	12.5	128,630	(1.1)	305,324	11.4
Georgia	43	57.3	0.0	0.0	1,036	X	1,962	(4.8)	2,988	X	984	X
India	61	20.5	1.3	57.2	169,569	0.5	11,424	1.2	68,173	1.2	48,136	(2.1)
Indonesia	23	64.5	28.5	53.8	31,146	19.9	11,800	1.2	111,516	(2.6)	26,695	(8.1)
Iran	39	3.3	0.0	0.0	18,500	21.7	44,000	0.0	11,400	0.0	88,300	(3.6)
Iraq	22	0.0	0.0	0.0	5,550	1.9	4,000	0.0	192	0.0	33,995	(0.3)
Israel	28	0.0	0.0	0.0	434	4.6	145	5.8	126	16.7	1,357	(3.2)
Japan	13	58.2	0.0	0.0	4,467	(7.0)	660	8.9	25,110	(0.2)	7,416	4.8
Jordan	13	0.0	0.0	0.0	405	17.1	791	0.0	70	5.5	7,627	(0.8)
Kazakhstan	83	22.9	2.9	100.0	35,239	X	186,549	X	9,600	X	35,685	X
Korea, North	17	38.7	0.0	0.0	2,007	2.7	50	0.0	7,370	0.0	2,614	(2.7)
Korea, South	22	16.5	0.0	0.0	2,053	(5.3)	91	30.8	6,450	(1.3)	1,270	(6.5)
Kuwait	8	0.0	0.0	0.0	5	114.3	137	2.2	2	0.0	1,638	(0.3)
Kyrgyzstan	54	14.0	0.0	0.0	1,387	X	8,900	X	730	X	8,163	X
Laos	7	30.0	2.1	100.0	900	18.5	800	0.0	12,560	(4.1)	8,820	4.7
Lebanon	31	0.7	0.0	0.0	306	2.7	14	40.0	80	(3.6)	623	(1.5)
Malaysia	24	63.8	14.5	48.5	7,535	46.6	281	8.9	22,248	0.0	2,790	(46.4)
Mongolia	76	49.6	8.2	0.0	1,357	4.4	117,983	(4.4)	13,750	(9.4)	23,560	40.5
Myanmar (Burma)	16	40.6	0.0	0.0	10,067	(0.1)	354	(1.8)	32,398	0.8	22,935	(1.0)
Nepal	31	22.4	0.0	0.0	2,556	10.0	1,757	(9.4)	5,750	4.7	4,237	(6.8)
Oman	5	0.0	0.0	0.0	63	44.3	1,000	0.0	X	X	20,183	(0.1)
Pakistan	34	5.8	0.0	0.0	21,323	4.7	5,000	0.0	3,477	15.1	47,288	(2.9)
Philippines	36	6.0	0.0	0.0	9,320	5.0	1,280	14.3	13,600	15.6	5,617	(30.3)
Saudi Arabia	58	0.0	0.0	0.0	3,777	67.6	120,000	38.5	1,800	33.3	89,392	(28.3)
Singapore	2	3.1	0.0	0.0	5	(83.3)	X	X	3	0.0	57	9.6
Sri Lanka	36	18.1	11.9	76.2	1,889	1.3	440	0.2	2,100	20.2	2,034	(15.6)
Syria	78	0.0	0.0	0.0	5,985	5.2	8,191	(1.8)	484	(2.4)	3,718	(3.6)
Tajikistan	31	4.2	4.9	100.0	846	X	3,533	X	537	X	9,144	X
Thailand	42	22.2	0.0	0.0	20,488	6.7	800	14.3	14,833	(3.7)	14,968	(5.1)
Turkey	52	11.3	0.0	0.0	27,611	2.1	2,378	22.2	20,199	0.0	16,775	(14.4)
Turkmenistan	67	4.1	0.0	0.0	1,471	X	33,202	X	4,000	X	8,320	X
United Arab Emirates	4	0.0	0.0	0.0	75	148.4	280	40.0	3	0.0	8,002	(1.5)
Uzbekistan	61	10.2	0.0	0.0	4,618	X	21,490	X	1,311	X	14,005	X
Vietnam	22	17.2	1.9	100.0	6,738	2.3	328	5.1	9,650	(3.9)	15,833	1.4
Yemen	33	0.0	0.0	0.0	1,520	3.8	16,065	0.0	2,000	(25.0)	33,212	1.9
EUROPE	**22**	**58.4**	**21.3**	**18.7**	**317,837**	**X**	**178,549**	**X**	**947,761**	**X**	**816,036**	**X**
Albania	41	37.3	0.0	0.0	702	(1.2)	424	5.6	1,049	1.7	565	(5.2)
Austria	43	52.8	0.0	0.0	1,506	(3.3)	1,985	(1.0)	3,233	(0.3)	1,550	5.4
Belarus	44	27.2	0.0	0.0	6,245	X	3,070	X	7,109	X	4,324	X
Belgium	45	21.0	0.0	0.0	790	2.4	687	(9.0)	709	2.0	1,095	3.4
Bosnia-Herzegovina	39	X	0.0	0.0	893	X	1,200	X	2,033	X	973	X
Bulgaria	54	31.7	0.0	0.0	4,286	3.5	1,811	(11.0)	3,348	0.1	1,609	5.0
Croatia	41	X	X	0.0	1,242	X	1,088	X	2,075	X	1,187	X

(Continued on next page)

| | Domestic and Forested Land | | | | Land Use (in Thousands of Hectares) | | | | | | | |
| | Domestic Land as a % of Land Area | Forests as a % of Original Forest | | Percent Frontier Forests Threatened | Cropland | | Permanent Pasture | | Forest and Woodland | | Other Land | |
COUNTRY	1994	Current Forests	Frontier Forests	1994	1992–94	% Change Since 1982–84	1992–94	% Change Since 1982–84	1992–94	% Change Since 1982–84	1992–94	% Change Since 1982–84
Czech Republic	55	X	X	0.0	2,265	X	588	X	173	X	547	X
Denmark	63	0.8	0.0	0.0	2,488	(5.3)	241	2.1	417	(13.8)	1,097	23.1
Estonia	34	29.4	0.0	0.0	1,145	X	311	X	2,018	X	753	X
Finland	9	82.3	1.1	100.0	2,585	4.5	112	(21.7)	23,186	(0.6)	4,575	1.2
France	55	16.5	0.0	0.0	19,387	1.7	10,830	(13.4)	14,938	2.3	9,854	11.5
Germany	50	26.3	0.0	0.0	11,885	(4.5)	5,255	(10.8)	10,700	4.0	7,087	12.5
Greece	68	17.0	0.0	0.0	3,502	(11.3)	5,252	(0.1)	2,620	0.0	1,516	42.2
Hungary	66	8.2	0.0	0.0	4,973	(6.1)	1,156	(9.4)	1,717	5.1	1,388	34.5
Iceland	23	X	0.0	0.0	6	(21.7)	2,274	0.0	120	0.0	7,625	0.0
Ireland	64	3.6	0.0	0.0	1,266	21.6	3,137	(32.8)	570	14.0	1,916	181.8
Italy	53	20.4	0.0	0.0	11,594	(5.8)	4,479	(11.2)	6,794	6.4	6,519	14.9
Latvia	41	19.8	0.0	0.0	1,720	X	808	X	2,841	X	836	X
Lithuania	54	16.0	0.0	0.0	3,041	X	472	X	1,983	X	984	X
Macedonia	51	X	0.0	0.0	662	X	638	X	1,035	X	227	X
Moldova	78	3.7	0.0	0.0	2,195	X	363	X	358	X	381	X
Netherlands	58	4.8	0.0	0.0	922	10.8	1,060	(10.0)	334	12.6	1,076	(0.9)
Norway	3	90.4	0.0	0.0	891	5.1	124	25.7	8,330	0.0	21,338	(0.3)
Poland	61	22.2	0.0	0.0	14,673	(1.0)	4,049	(0.7)	8,732	0.7	2,988	3.9
Portugal	43	9.4	0.0	0.0	3,057	(3.0)	900	7.4	3,102	4.6	2,091	(4.7)
Romania	64	41.5	0.0	0.0	9,942	(5.8)	4,851	9.6	6,681	1.8	1,559	4.6
Russian Federation	13	68.7	29.3	18.6	133,072	X	86,858	X	767,347	X	701,573	X
Slovak Republic	51	X	0.0	0.0	1,074	X	557	X	1,326	X	249	X
Slovenia	39	X	0.0	0.0	296	X	540	X	1,077	X	98	X
Spain	62	15.1	0.0	0.0	19,910	(2.9)	10,416	0.5	16,063	3.1	3,555	1.7
Sweden	8	86.0	2.9	100.0	2,776	(5.6)	576	(17.0)	28,025	0.1	9,653	1.3
Switzerland	40	44.8	0.0	0.0	433	5.1	1,148	(28.6)	1,186	12.7	1,187	34.6
Ukraine	72	20.4	0.0	0.0	34,410	X	7,483	X	9,239	X	6,803	X
United Kingdom	71	6.0	0.0	0.0	6,224	(10.9)	11,090	(1.4)	2,390	8.2	4,456	19.7
Yugoslavia (Serbia-Montenegro)	61	X	0.0	0.0	4,080	X	2,119	X	1,769	X	2,233	X
OCEANIA	**57**	**64.9**	**22.3**	**76.3**	**51,553**	**1.4**	**430,077**	**(2.8)**	**200,252**	**(0.2)**	**164,807**	**6.3**
Australia	60	64.3	17.8	62.8	47,023	1.7	415,700	(2.8)	145,000	(0.4)	158,057	6.4
Fiji	24	49.9	0.0	0.0	260	40.5	174	29.1	1,185	0.0	208	(35.5)
New Zealand	62	29.2	8.9	100.0	3,307	(5.5)	13,774	(2.2)	7,667	2.6	2,051	17.2
Papua New Guinea	1	85.4	39.6	83.5	423	13.4	90	(4.6)	42,000	0.0	2,776	(1.5)
Solomon Islands	3	93.9	0.0	0.0	57	6.9	39	0.0	2,450	(2.8)	253	35.5

Source: World Resources Institute, *World Resources 1998–1999, 1998.*

Table K

World Countries: Infrastructure, 1994

COUNTRY	ELECTRIC POWER		TELECOMMUNICATIONS		PAVED ROADS		WATER		RAILWAYS	
	Production (Kwh/ Person)	System Losses (% of Total Output)	Telephone Mainlines (Per 100 Persons)	Faults (Per 100 Mainlines Per Year)	Road Density (Km Per Million Persons)	Roads in Good Condition (% of Paved Roads)	Population with Access to Safe Water (% of Total)	Losses (% of Total Water Provision)	Rail Traffic Units (Per 1000 $GDP)	Diesels in Use (% of Diesel Inventory)
AFRICA										
Algeria	701	15	37	82	2,043	40	X	X	76	88
Angola	X	X	X	X	X	X	X	X	X	X
Benin	44	1	3	150	241	26	50	X	X	X
Botswana	X	X	27	55	1,977	94	X	25	X	X
Burkina Faso	X	X	2	X	158	24	67	X	X	X
Burundi	X	X	2	81	177	58	38	46	X	X
Cameroon	223	4	5	74	299	38	34	X	96	72
Central African Republic	X	X	2	X	135	30	12	X	X	X
Chad	X	X	1	152	56	X	X	X	X	X
Congo	X	X	X	X	X	X	X	X	X	X
Congo Republic	181	0	7	54	509	50	X	X	140	31
Cote d'Ivoire	144	4	7	80	357	75	83	16	32	44
Egypt	849	12	39	X	633	39	41	X	465	75
Equatorial Guinea	X	X	X	X	X	X	X	X	X	X
Eritrea	25	X	X	X	X	X	X	X	X	60
Ethiopia	X	3	3	74	77	48	18	46	X	X
Gabon	928	11	23	74	511	30	72	22	90	X
Gambia	X	X	14	120	772	X	77	X	X	X
Ghana	366	2	3	159	474	28	56	47	28	X
Guinea	X	X	2	4	229	27	33	X	X	X
Guinea-Bissau	X	X	6	X	X	X	25	X	X	X
Kenya	X	X	X	X	X	X	X	X	X	X
Lesotho	X	X	6	6	452	53	46	X	X	X
Liberia	X	X	X	X	X	X	X	X	X	X
Libya	X	X	X	X	X	X	X	X	X	X
Madagascar	X	X	3	78	433	56	53	X	X	X
Malawi	X	X	3	X	278	56	49	X	26	70
Mali	X	X	1	165	308	63	66	X	104	48
Mauritania	X	X	3	67	804	58	100	X	X	X
Mauritius	383	3	72	84	1,549	95	X	X	X	81
Morocco	X	X	25	10	179	20	22	5	125	X
Mozambique	24	24	3	78	343	12	12	X	X	X
Namibia	X	X	40	79	2,722	60	59	X	17	18
Niger	137	39	1	327	400	67	42	X	X	X
Nigeria	X	X	3	38	376	41	64	X	75	68
Rwanda	X	X	2	36	162	28	51	X	X	X
Senegal	99	9	8	17	542	62	43	X	X	X
Sierra Leone	X	X	3	X	295	X	X	X	X	X
Somalia	X	X	X	X	X	X	X	X	X	X
South Africa	4,329	7	89	X	1,394	X	X	X	804	82
Sudan	X	X	X	X	X	X	X	X	X	X
Swaziland	X	X	X	X	X	X	X	X	X	X
Tanzania	66	12	3	X	142	25	52	X	X	50

(Continued on next page)

COUNTRY	ELECTRIC POWER		TELECOMMUNICATIONS		PAVED ROADS		WATER		RAILWAYS	
	Production (Kwh/Person)	System Losses (% of Total Output)	Telephone Mainlines (Per 100 Persons)	Faults (Per 100 Mainlines Per Year)	Road Density (Km Per Million Persons)	Roads in Good Condition (% of Paved Roads)	Population with Access to Safe Water (% of Total)	Losses (% of Total Water Provision)	Rail Traffic Units (Per 1000 $GDP)	Diesels in Use (% of Diesel Inventory)
Togo	X	X	4	22	470	40	71	X	X	X
Tunisia	731	6	45	113	2,080	55	67	30	119	57
Uganda	X	X	2	58	118	10	15	X	20	67
Zambia	900	11	9	33	795	40	59	X	169	44
Zimbabwe	790	7	12	215	1,406	27	36	X	523	83
NORTH AND CENTRAL AMERICA										
Belize	X	X	X	X	X	X	X	X	X	X
Canada	18,309	7	592	X	11,541	X	100	X	325	91
Costa Rica	X	X	102	X	1,756	22	94	X	X	46
Cuba	X	X	X	X	X	X	X	X	X	X
Dominican Republic	X	X	66	133	364	52	62	X	X	X
El Salvador	X	X	31	58	323	7	41	X	X	X
Guatemala	290	15	22	X	320	X	60	X	X	X
Haiti	X	X	X	X	X	X	X	X	X	X
Honduras	X	X	21	40	443	50	X	X	X	X
Jamaica	897	20	70	84	1,881	10	72	31	X	X
Mexico	1,381	14	80	X	1,019	85	78	X	73	75
Nicaragua	X	X	14	X	414	X	53	20	X	X
Panama	1,167	24	97	10	1,332	36	83	X	X	X
Trinidad and Tobago	3,122	13	142	6	1,724	72	96	X	X	X
United States	12,900	8	565	X	14,455	X	X	X	344	90
SOUTH AMERICA										
Argentina	1,670	15	123	13	1,856	35	64	X	120	68
Bolivia	349	14	33	28	258	21	46	X	81	62
Brazil	1,580	15	71	43	929	30	96	30	61	62
Chile	1,646	11	94	82	808	42	86	X	42	57
Colombia	1,032	18	85	83	383	42	X	38	5	35
Ecuador	675	24	48	197	476	53	58	47	X	X
Guyana	X	X	X	X	X	X	X	X	X	X
Paraguay	6,693	0	28	X	X	X	33	X	X	X
Peru	587	11	27	47	347	24	58	X	16	X
Suriname	X	X	X	X	X	X	X	X	X	X
Uruguay	2,842	14	168	16	2,106	26	X	X	13	62
Venezuela	3,404	15	91	5	10,269	40	89	X	X	X
ASIA										
Afghanistan	X	X	X	X	X	X	X	X	X	X
Armenia	1,850	22	157	X	2,024	X	X	X	X	X
Azerbaijan	2,699	13	89	X	X	X	X	X	X	X
Bangladesh	79	32	2	X	59	15	78	47	37	74
Bhutan	X	X	X	X	X	X	X	X	X	X
Cambodia	X	X	X	X	X	X	X	X	X	X
China	647	7	10	X	X	X	71	X	847	82
Georgia	2,120	23	105	43	X	X	X	X	X	X

(Continued on next page)

COUNTRY	ELECTRIC POWER		TELECOMMUNICATIONS		PAVED ROADS		WATER		RAILWAYS	
	Production (Kwh/Person)	System Losses (% of Total Output)	Telephone Mainlines (Per 100 Persons)	Faults (Per 100 Mainlines Per Year)	Road Density (Km Per Million Persons)	Roads in Good Condition (% of Paved Roads)	Population with Access to Safe Water (% of Total)	Losses (% of Total Water Provision)	Rail Traffic Units (Per 1000 $GDP)	Diesels in Use (% of Diesel Inventory)
India	373	23	8	218	893	20	75	X	488	90
Indonesia	233	17	8	49	160	30	42	29	27	75
Iran	1,101	12	50	X	X	X	89	X	61	39
Iraq	X	X	X	X	X	X	X	X	X	X
Israel	4,870	3	353	21	2,658	X	100	X	26	82
Japan	7,211	4	464	2	6,426	X	X	X	147	88
Jordan	1,120	13	71	89	1,767	X	99	41	74	76
Kazakhstan	4,826	9	88	X	6,747	X	X	X	X	54
Korea, North	X	X	X	X	X	X	X	X	X	X
Korea, South	2,996	5	357	13	1,090	70	78	X	146	88
Kuwait	8,924	9	245	30	X	X	100	X	X	X
Kyrgyzstan	2,636	10	75	30	X	X	X	X	X	X
Laos	X	X	2	12	516	X	28	X	X	X
Lebanon	X	X	X	X	X	X	X	X	X	X
Malaysia	1,612	9	112	78	X	X	78	29	30	76
Mongolia	X	X	30	43	X	X	66	X	X	58
Myanmar (Burma)	61	35	2	X	210	40	33	X	X	75
Nepal	45	24	3	168	139	66	37	45	X	X
Oman	2,729	1	74	24	2,992	18	57	40	X	78
Pakistan	435	17	10	120	826	18	50	40	137	X
Philippines	419	13	10	10	242	31	81	53	X	X
Saudi Arabia	4,417	9	93	24	3,601	X	95	X	X	90
Singapore	6,353	5	415	11	993	X	100	8	65	X
Sri Lanka	200	17	8	X	536	10	60	X	X	X
Syria	X	X	X	X	X	X	X	X	X	X
Tajikistan	3,001	7	48	218	X	X	X	X	X	X
Thailand	1,000	10	31	32	841	50	72	48	75	72
Turkey	1,154	13	160	27	5,514	X	91	44	65	76
Turkmenistan	3,422	11	65	53	X	X	X	X	X	X
United Arab Emirates	9,917	6	321	X	2,706	X	100	X	X	X
Uzbekistan	2,390	10	67	X	X	X	X	X	X	X
Vietnam	139	24	2	X	X	X	50	X	X	60
Yemen	156	11	11	22	372	39	X	45	X	X
EUROPE										
Albania	1,002	13	13	28	414	X	100	X	X	78
Austria	6,554	6	440	35	13,954	X	100	X	213	83
Belarus	3,692	11	169	X	4,707	X	X	X	X	92
Belgium	7,215	5	425	8	12,909	X	100	X	120	83
Bosnia-Herzegovina	X	X	X	X	X	X	X	X	X	X
Bulgaria	4,000	14	275	48	3,986	X	100	X	297	78
Croatia	X	X	X	X	X	X	X	X	X	X
Czech Republic	5,740	7	176	X	13,741	X	X	X	X	X
Denmark	5,983	7	581	X	X	X	100	X	89	X
Estonia	7,599	9	215	45	5,180	X	X	X	X	77
Finland	11,409	5	544	11	9,429	X	100	X	180	99
France	8,089	6	525	8	13,008	X	100	X	140	93
Germany	6,693	2	457	14	X	X	100	X	107	88

(Continued on next page)

COUNTRY	ELECTRIC POWER		TELECOMMUNICATIONS		PAVED ROADS		WATER		RAILWAYS	
	Production (Kwh/Person)	System Losses (% of Total Output)	Telephone Mainlines (Per 100 Persons)	Faults (Per 100 Mainlines Per Year)	Road Density (Km Per Million Persons)	Roads in Good Condition (% of Paved Roads)	Population with Access to Safe Water (% of Total)	Losses (% of Total Water Provision)	Rail Traffic Units (Per 1000 $GDP)	Diesels in Use (% of Diesel Inventory)
Greece	3,624	7	487	80	10,341	X	100	X	37	47
Hungary	3,080	10	125	60	7,756	X	100	X	369	78
Iceland	X	X	X	X	X	X	X	X	X	X
Ireland	4,545	9	314	38	24,468	X	100	X	54	60
Italy	3,963	7	410	17	5,283	X	100	X	96	79
Latvia	1,460	26	247	26	4,437	X	X	X	X	93
Luthuania	5,050	9	222	46	9,529	X	X	X	X	64
Macedonia	2,812	8	148	13	2,310	X	X	X	X	X
Moldova	2,562	11	117	45	2,832	X	X	X	X	X
Netherlands	5,089	4	487	4	6,078	X	100	X	90	85
Norway	27,501	8	529	16	14,698	X	100	X	X	X
Poland	3,473	11	103	X	6,132	69	100	X	610	67
Portugal	3,055	11	306	52	6,130	50	100	X	97	86
Romania	2,386	10	113	116	3,431	30	100	28	X	52
Russian Federation	6,820	8	154	X	X	X	X	X	X	X
Slovak Republic	4,251	8	167	23	X	X	77	X	X	X
Slovenia	6,238	5	247	6	5,525	X	X	X	X	81
Spain	4,022	9	353	6	8,540	X	100	X	67	87
Sweden	16,913	6	682	10	11,747	X	100	X	201	88
Switzerland	8,471	6	606	21	10,299	X	100	X	X	X
Ukraine	4,900	9	145	49	3,085	X	X	X	X	60
United Kingdom	5,660	8	473	16	6,224	X	100	X	64	74
Yugoslavia (Serbia-Montenegro)	X	X	X	X	X	X	X	X	X	X
OCEANIA										
Australlia	9,221	7	471	X	16,221	X	100	X	75	81
Fiji	X	X	X	X	X	X	X	X	X	X
New Zealand	9,086	8	449	X	15,725	X	97	X	64	90
Papua New Guinea	X	X	9	X	196	34	33	X	X	X
Solomon Islands	X	X	X	X	X	X	X	X	X	X

Source: World Bank, *Human Development Report,* 1995.

Electric power as production in kwh/person and system losses as percent of total output; telecommunications in telephone mainlines per 100 persons and faults per 100 mainlines; paved roads as density (km per million persons) and good condition (percent of paved roads); water as population with access to safe water (percent of total population) and losses as percent of total water; and railways as rail traffic units per $1000 GDP and diesels in use as percent of locomotive inventory.

Table L

World Countries: Globally Threatened Plant and Animal Species—1996

COUNTRY	Mammals Threatened Species	Mammals Number of Species per 000 km²	Birds Threatened Species	Birds Number of Species per 000 km²	Reptiles Threatened Species	Reptiles Number of Species per 000 km²	Amphibians Threatened Species	Amphibians Number of Species per 000 km²	Fresh-water Fish Threatened Species	Plants Rare and Threatened Species	Plants Number of Species per 000 km²
AFRICA											
Algeria	15	15	8	32	1	X	0	X	1	145	509
Angola	17	56	13	156	5	X	0	X	0	25	1,017
Benin	9	85	1	138	2	X	0	X	0	3	899
Botswana	5	43	7	101	0	41	0	X	0	4	X
Burkina Faso	6	49	1	112	1	X	0	10	0	0	369
Burundi	5	76	6	322	0	X	0	X	0	1	1,783
Cameroon	32	83	14	193	3	X	1	X	26	74	2,237
Central African Republic	11	53	2	137	1	X	0	X	0	0	921
Chad	14	27	3	75	1	X	0	X	0	12	322
Congo (Zaire)	38	69	26	153	3	X	0	X	1	7	1,817
Congo Rep.	10	62	3	140	2	X	0	X	0	3	1,356
Cote d'Ivoire	16	73	12	170	4	X	1	X	0	66	1,118
Egypt	15	21	11	33	6	18	0	1	0	84	452
Equatorial Guinea	12	131	4	194	2	X	1	X	0	9	2,135
Eritrea	6	49	3	140	3	X	0	X	0	X	X
Ethiopia	35	54	20	133	1	X	0	X	0	153	1,378
Gabon	12	64	4	157	3	X	0	X	0	78	2,197
Gambia	4	104	1	269	1	X	0	X	0	0	928
Ghana	13	78	10	186	4	X	1	X	0	32	1,264
Guinea	11	66	12	142	3	X	1	X	0	35	1,043
Guinea-Bissau	4	71	1	159	3	X	0	X	0	0	655
Kenya	43	94	24	221	5	49	0	23	20	158	1,571
Lesotho	2	23	5	40	0	X	0	X	1	7	1,093
Liberia	11	87	13	168	3	28	1	17	0	1	1,037
Libya	11	14	2	17	3	X	0	X	0	57	327
Madagascar	46	27	28	53	17	66	2	38	13	189	2,347
Malawi	7	86	9	230	0	55	0	31	0	61	1,592
Mali	13	28	6	81	1	3	0	X	0	14	355
Mauritania	14	13	3	59	3	X	0	X	0	3	239
Mauritius	4	7	10	46	6	19	0	0	0	222	1,183
Morocco	18	30	11	60	2	X	0	X	1	195	1,028
Mozambique	13	42	14	117	5	X	0	15	2	92	1,294
Namibia	11	36	8	109	3	X	1	7	3	23	729
Niger	11	27	2	60	1	X	0	X	0	0	237
Nigeria	26	62	9	153	4	X	0	X	0	9	1,036
Rwanda	9	110	6	373	0	X	0	X	0	0	1,662
Senegal	13	58	6	144	7	X	0	X	0	32	771
Sierra Leone	9	77	12	243	3	X	0	X	0	12	1,091
Somalia	18	43	8	107	2	49	0	7	3	57	761
South Africa	33	51	16	122	19	61	9	19	27	953	4,711
Sudan	21	43	9	110	3	X	0	X	0	8	506
Swaziland	5	37	6	303	0	85	0	33	0	41	2,197
Tanzania	33	70	30	183	4	63	0	28	19	406	229
Togo	8	110	1	220	3	X	0	X	0	0	1,128
Tunisia	11	31	6	69	2	X	0	X	0	24	855

(Continued on next page)

COUNTRY	Mammals		Birds		Reptiles		Amphibians		Fresh-water Fish	Plants	
	Threatened Species	Number of Species per 000 km²	Threatened Species	Number of Species per 000 km²	Threatened Species	Number of Species per 000 km²	Threatened Species	Number of Species per 000 km²	Threatened Species	Rare and Threatened Species	Number of Species per 000 km²
Uganda	18	118	10	290	1	52	0	17	28	6	1,762
Zambia	11	55	10	145	0	X	0	20	0	9	1,105
Zimbabwe	9	81	9	159	0	46	0	36	0	94	1,253
NORTH AMERICA											
Canada	7	20	5	44	3	4	1	4	13	649	299
United States	35	45	50	68	28	29	24	24	123	1,845	1,679
CENTRAL AMERICA											
Belize	5	95	1	271	5	81	0	24	0	41	2,090
Costa Rica	14	120	13	350	7	125	1	95	0	456	6,421
Cuba	9	14	13	62	7	46	0	19	4	811	2,714
Dominican Republic	4	12	11	81	10	62	1	21	0	73	2,965
El Salvador	2	106	0	196	6	57	0	18	0	35	1,956
Guatemala	8	114	4	208	9	105	0	45	0	315	3,638
Haiti	4	2	11	54	6	73	1	33	0	28	3,345
Honduras	7	78	4	190	7	68	0	25	0	55	2,252
Jamaica	4	23	7	110	8	35	4	20	0	371	2,662
Mexico	64	79	36	135	18	120	3	50	86	1,048	4,382
Nicaragua	4	86	3	207	7	69	0	25	0	78	3,003
Panama	17	112	10	376	7	116	0	84	1	561	4,618
Trinidad and Tobago	1	125	3	324	5	87	0	32	0	16	2,470
SOUTH AMERICA											
Argentina	27	50	41	140	5	34	5	23	1	170	1,407
Bolivia	24	67	27	X	3	44	0	24	0	49	3,500
Brazil	71	43	103	161	15	51	5	54	12	463	5,935
Chile	16	22	18	71	1	17	3	10	4	292	1,229
Colombia	35	75	64	355	15	122	0	123	5	376	10,479
Ecuador	28	100	53	460	12	124	0	133	1	375	6,052
Guyana	10	70	3	246	8	X	0	X	0	47	2,180
Paraguay	10	90	26	164	3	35	0	25	0	12	2,208
Peru	46	69	64	310	9	60	1	63	0	377	3,448
Suriname	10	72	2	240	6	60	0	38	0	48	1,870
Uruguay	5	31	11	92	0	X	0	X	0	11	845
Venezuela	24	69	22	266	14	58	0	45	5	107	4,510
ASIA											
Afghanistan	11	31	13	59	1	26	1	2	0	6	882
Armenia	4	X	5	X	3	32	0	4	0	0	X
Azerbaijan	11	X	8	X	3	26	0	4	5	1	X
Bangladesh	18	45	30	122	13	49	0	8	0	24	2,074
Bhutan	20	59	14	269	1	11	0	14	0	20	3,268
Cambodia	23	47	18	118	9	32	0	11	5	7	X
China	75	41	90	114	15	35	1	27	28	343	3,112
Georgia	10	X	5	X	7	24	0	6	3	1	X
India	75	47	73	136	16	57	3	29	4	1,256	2,216

(Continued on next page)

COUNTRY	Mammals Threatened Species	Mammals Number of Species per 000 km²	Birds Threatened Species	Birds Number of Species per 000 km²	Reptiles Threatened Species	Reptiles Number of Species per 000 km²	Amphibians Threatened Species	Amphibians Number of Species per 000 km²	Fresh-water Fish Threatened Species	Plants Rare and Threatened Species	Plants Number of Species per 000 km²
Indonesia	128	77	104	269	19	90	0	48	60	281	4,864
Iran	20	26	14	60	8	30	2	2	7	1	X
Iraq	7	23	12	49	2	23	0	2	2	2	X
Israel	13	72	8	141	5	X	0	X	2	38	X
Japan	29	40	33	X	8	20	10	16	0	704	1,418
Jordan	7	34	4	68	1	X	0	X	7	10	1,069
Kazakhstan	15	X	15	X	1	6	1	2	0	0	X
Korea, North	7	X	19	51	0	8	0	6	5	7	1,274
Korea, South	6	23	19	53	0	12	0	7	0	69	1,360
Kuwait	1	17	3	17	2	24	0	2	0	0	193
Kyrgyzstan	6	X	5	X	1	9	0	1	0	1	X
Laos	30	61	27	171	7	23	0	13	4	5	X
Lebanon	5	53	5	152	2	X	0	X	4	4	X
Malaysia	42	90	34	158	14	85	0	50	14	510	4,732
Mongolia	12	25	14	X	0	4	0	2	0	1	429
Myanmar (Burma)	31	62	44	216	20	51	0	19	1	29	1,742
Nepal	28	70	27	255	5	33	0	15	0	21	2,716
Oman	9	20	5	39	4	23	0	X	3	4	371
Pakistan	13	36	25	88	6	41	0	4	1	12	1,163
Philippines	49	50	86	129	7	62	2	21	26	371	2,604
Saudi Arabia	9	13	11	26	2	14	0	X	0	6	294
Singapore	6	113	9	295	1	X	0	X	1	14	5,007
Sri Lanka	14	47	11	134	8	77	0	21	8	436	1,613
Syria	4	24	7	78	3	X	0	X	0	10	X
Tajikistan	5	X	9	X	1	16	0	1	1	0	X
Thailand	34	72	45	168	16	81	0	29	14	382	2,999
Turkey	15	28	14	72	12	24	2	4	18	1,827	2,012
Turkmenistan	11	X	12	X	2	22	0	1	5	1	X
United Arab Emirates	3	12	4	33	2	18	0	X	1	0	X
Uzbekistan	7	X	11	X	0	15	0	1	3	5	X
Vietnam	38	67	47	168	12	57	1	25	3	350	X
Yemen	5	18	13	39	2	21	0	X	0	X	X
EUROPE											
Albania	2	48	7	162	1	22	0	9	7	50	2,093
Austria	7	41	5	106	1	7	0	10	7	22	1,462
Belarus	4	X	4	81	0	3	0	4	0	0	X
Belgium	6	40	3	125	0	6	0	12	1	3	969
Bosnia-Herzegovina	10	X	2	X	X	X	1	X	6	0	X
Bulgaria	13	37	12	108	1	15	0	8	8	94	1,584
Croatia	10	X	4	126	X	X	1	X	20	0	X
Czech Republic	7	X	6	101	X	X	X	X	6	X	X
Denmark	3	27	2	121	0	3	0	9	0	6	741
Estonia	4	40	2	130	0	3	0	7	1	2	992
Finland	4	18	4	78	0	2	0	2	1	11	325
France	13	25	7	72	3	9	2	9	3	117	1,198
Germany	8	23	5	73	0	4	0	6	7	X	X
Greece	13	41	10	107	6	22	1	6	16	539	2,091
Hungary	8	34	10	98	1	7	0	8	11	24	1,029

(Continued on next page)

COUNTRY	Mammals Threatened Species	Mammals Number of Species per 000 km²	Birds Threatened Species	Birds Number of Species per 000 km²	Reptiles Threatened Species	Reptiles Number of Species per 000 km²	Amphibians Threatened Species	Amphibians Number of Species per 000 km²	Fresh-water Fish Threatened Species	Plants Rare and Threatened Species	Plants Number of Species per 000 km²
Iceland	1	5	0	41	0	0	0	0	0	1	157
Ireland	2	13	1	75	0	1	0	2	1	9	469
Italy	10	29	7	76	4	13	4	11	9	273	1,776
Latvia	4	45	6	117	0	4	0	7	1	0	623
Lithuania	5	37	4	109	0	4	0	7	1	0	646
Macedonia	10	X	3	X	1	X	X	X	4	X	X
Moldova	2	46	7	119	1	6	0	9	9	1	X
Netherlands	6	35	3	120	0	4	0	10	1	1	758
Norway	4	17	3	77	0	2	0	2	1	20	524
Poland	10	27	6	72	0	3	0	6	2	27	738
Portugal	13	30	7	99	0	14	1	8	9	240	1,200
Romania	16	29	11	87	2	9	0	7	11	122	1,116
Russian Federation	31	23	38	54	5	5	0	2	13	127	X
Slovak Republic	8	X	4	124	0	X	0	X	7	X	X
Slovenia	10	55	3	164	0	17	1	X	5	11	X
Spain	19	22	10	76	6	15	3	7	10	896	X
Sweden	5	17	4	71	0	2	0	4	1	19	1,400
Switzerland	6	47	4	121	0	9	0	11	4	9	1,033
Ukraine	15	X	10	85	2	6	0	5	12	16	756
United Kingdom	4	17	2	80	0	3	0	2	1	28	539
Yugoslavia (Serbia-Montenegro)	12	X	8	X	1	X	X	X	13	X	X
OCEANIA											
Australia	58	28	45	72	37	83	25	23	37	1,597	1,672
Fiji	4	3	9	61	6	20	1	2	0	72	1,071
New Zealand	3	3	44	51	11	13	1	1	8	236	727
Papua New Guinea	57	60	31	182	10	79	0	56	13	95	2,821
Solomon Islands	20	37	18	115	4	43	0	12	0	43	1,959

Source: World Conservation Monitoring Centre and World Conservation Union; World Resources Institute, *World Resources 1998–99, 1998.*

Table M

World Countries: Water Resources

| COUNTRY | Annual Renewable Water Resources[a] | | Annual Average Groundwater Resources[b] | | Sectoral Withdrawals (%)[c] | | | | | |
| | Supply Per Capita (cubic meters) 1998 est. | Withdrawal Per Capita (cubic meters) 1998 | Recharge Per Capita (cubic meters) 1998 est. | Withdrawal Per Capita (cubic meters) 1998 | Domestic | | Industry | | Agriculture | |
					Surface	Ground	Surface	Ground	Surface	Ground
WORLD	**6,918**	**645**	**X**	**X**	**8**	**X**	**23**	**X**	**69**	**X**
AFRICA	**5,133**	**202**	**X**	**X**	**7**	**X**	**5**	**X**	**88**	**X**
Algeria	460	180	56	91	25	X	15	X	60	X
Angola	15,376	57	6,017	X	14	X	10	X	76	X
Benin	1,751	28	306	X	23	X	10	X	67	X
Botswana	1,870	84	1,096	X	32	X	20	X	48	X
Burkina Faso	1,535	39	833	X	19	X	0	X	81	X
Burundi	546	20	319	X	36	X	0	X	64	X
Cameroon	18,711	38	6,982	X	46	X	19	X	35	X
Central African Republic	40,413	26	16,050	X	21	X	5	X	74	X
Chad	2,176	34	1,669	15.3	16	29.4	2	0.0	82	70.6
Congo (Zaire)	19,001	10	149,185	X	61	X	16	X	23	X
Congo Republic	78,668	20	70,163	X	62	X	27	X	11	10.00
Cote d'Ivoire	5,265	67	2,588	X	22	X	11	X	67	X
Egypt	43	921	20	68.3	6	X	8	X	86	X
Equatorial Guinea	69,767	15	24,256	X	81	X	13	X	6	X
Eritrea	789	X	X	X	X	X	X	X	X	X
Ethiopia	1,771	51	708	X	11	X	3	X	86	X
Gabon	140,171	70	52,991	0.36	72	100.0	22	0.0	6	0.0
Gambia	2,513	29	419	X	7	X	2	X	91	X
Ghana	1,607	35	1,326	X	35	X	13	X	52	X
Guinea	29,454	142	4,952	X	10	X	3	X	87	X
Guinea-Bissau	14,109	17	142,346	X	60	X	4	X	36	X
Kenya	696	87	103	X	20	X	4	X	76	X
Lesotho	2,395	30	229	X	22	X	22	X	56	X
Liberia	72,780	54	21,834	X	27	X	13	X	60	X
Libya	100	880	84	554.7	11	13.3	2	4.3	87	82.5
Madagascar	20,614	1,579	3,364	461.2	1	100.0	0	0.0	99	0.0
Malawi	1,690	98	135	X	10	X	3	X	86	X
Mali	5,071	162	1,690	11.2	2	7.1	1	0.0	97	92.9
Mauritania	163	923	122	498.3	6	X	2	X	92	X
Mauritius	1,915	410	589	X	16	X	7	X	77	X
Morocco	1,071	433	268	138.6	5	X	3	X	92	X
Mozambique	5,350	40	910	X	9	X	2	X	89	X
Namibia	3,751	179	1,270	X	29	X	3	X	68	X
Niger	346	69	247	17.9	16	57.7	2	3.8	82	38.5
Nigeria	1,815	41	714	X	31	X	15	X	54	X
Rwanda	965	135	551	X	5	X	2	X	94	X
Senegal	2,933	202	844	39.2	5	25.0	3	0.0	92	75.0

(Continued on next page)

COUNTRY	Annual Renewable Water Resources[a]		Annual Average Groundwater Resources[b]		Sectoral Withdrawals (%)[c]					
					Domestic		Industry		Agriculture	
	Supply Per Capita (cubic meters) 1998 est.	Withdrawal Per Capita (cubic meters) 1998	Recharge Per Capita (cubic meters) 1998 est.	Withdrawal Per Capita (cubic meters) 1998	Surface	Ground	Surface	Ground	Surface	Ground
Sierra Leone	34,957	98	10,924	X	7	X	4	X	89	X
Somalia	563	99	310	38.1	3	X	0	X	97	X
South Africa	1,011	359	108	61.4	17	10.6	11	5.6	72	83.8
Sudan	1,227	666	245	13.0	4	X	1	X	94	X
Swaziland	2,836	1,171	X	X	2	X	2	X	96	X
Tanzania	2,485	40	932	X	9	X	2	X	89	X
Togo	2,594	28	1,286	X	62	X	16	X	25	X
Tunisia	371	376	127	167.7	9	13.6	6	0.0	89	86.4
Uganda	1,829	20	1,360	X	32	X	8	X	60	X
Zambia	9,229	216	5,420	X	16	X	7	X	77	X
Zimbabwe	1,182	136	419	X	14	X	7	X	79	X
NORTH AMERICA	17,458	1,798	X	X	13	X	47	X	39	X
Canada	94,373	1,602	12,241	38.8	18	43.3	70	14.2	12	42.5
United States	8,983	1,839	5,531	432.9	13	22.7	45	6.1	42	71.1
CENTRAL AMERICA	8,084	916	X	X	6	X	8	X	86	X
Belize	69,565	109	X	X	10	X	0	X	90	X
Costa Rica	26,027	780	5,753	X	4	X	7	X	89	X
Cuba	3,104	870	720	408.3	9	X	2	X	89	X
Dominican Republic	2,430	446	364	X	5	X	6	X	89	X
El Salvador	3,128	244	X	X	7	X	4	X	89	X
Guatemala	10,033	139	2,681	X	9	X	17	X	74	X
Haiti	1,460	7	382	X	24	X	8	X	68	X
Honduras	9,015	294	6,345	X	4	X	5	X	91	X
Jamaica	3,269	159	X	X	7	X	7	X	86	X
Mexico	3,729	915	1,450	311.4	6	13.2	8	23.0	86	63.8
Nicaragua	39,203	368	13,217	X	25	X	21	X	54	X
Panama	52,042	754	15,179	X	12	X	11	X	77	X
Trinidad and Tobago	3,869	148	X	X	27	X	38	X	35	X
SOUTH AMERICA	28,072	335	X	X	18	X	23	X	59	X
Argentina	19,212	1,043	3,543	180.4	9	10.6	18	19.1	73	70.2
Bolivia	37,703	201	16,338	X	10	X	5	X	85	X
Brazil	31,424	246	11,347	X	22	X	19	X	59	X
Chile	31,570	1,625	9,444	X	6	X	5	X	89	X
Colombia	28,393	174	13,533	X	41	X	16	X	43	X
Ecuador	25,791	581	11,006	X	7	X	3	X	90	X
Guyana	281,542	1,819	120,327	X	1	X	0	X	99	X
Paraguay	18,001	112	7,851	X	15	X	7	X	78	X
Peru	1,613	300	12,219	139.4	19	25.0	9	15.0	72	60.0
Suriname	452,489	1,192	180,995	X	6	X	5	X	89	X

(Continued on next page)

| COUNTRY | Annual Renewable Water Resources[a] | | Annual Average Groundwater Resources[b] | | Sectoral Withdrawals (%)[c] | | | | | |
	Supply Per Capita (cubic meters) 1998 est.	Withdrawal Per Capita (cubic meters) 1998	Recharge Per Capita (cubic meters) 1998 est.	Withdrawal Per Capita (cubic meters) 1998	Domestic Surface	Domestic Ground	Industry Surface	Industry Ground	Agriculture Surface	Agriculture Ground
Uruguay	18,215	241	7,101	X	6	X	3	X	91	X
Venezuela	36,830	382	9,767	X	43	X	11	X	46	X
ASIA	**3,680**	**542**	**X**	**X**	**6**	**X**	**9**	**X**	**85**	**X**
Afghanistan	2,354	1,825	1,241	X	1	X	0	X	99	X
Armenia	2,493	804	1,154	X	30	X	4	X	66	X
Azerbaijan	1,069	2,177	856	X	5	X	25	X	70	X
Bangladesh	10,940	217	274	39.6	3	12.9	1	0.9	96	86.2
Bhutan	49,557	13	X	X	36	X	10	X	54	X
Cambodia	8,195	66	2,790	X	5	X	1	X	94	X
China	2,231	461	693	69.7	6	X	7	X	87	53.6
Georgia	10,682	637	3,166	X	21	X	20	X	59	X
India	1,896	612	359	222.4	3	3.1	4	1.3	93	95.7
Indonesia	12,251	96	1,094	X	13	X	11	X	76	X
Iran	1,755	1,079	671	738.8	6	X	2	X	92	X
Iraq	1,615	2,368	55	13.1	3	55.6	5	44.4	92	0.0
Israel	289	407	187	279.1	16	X	5	X	79	X
Japan	4,344	735	1,469	104.3	17	29.3	33	40.7	50	30.1
Jordan	114	201	97	107.0	22	X	3	X	75	X
Kazakhstan	4,484	2,002	2,133	X	2	X	17	X	81	X
Korea, North	2,887	727	X	X	11	X	16	X	73	X
Korea, South	1,434	632	X	29.4	19	0.0	35	83.3	46	16.7
Kuwait	11	307	X	X	37	X	2	X	60	X
Kyrgyzstan	10,394	2,257	3,043	X	3	X	3	X	94	X
Laos	50,392	259	9,332	X	8	X	10	X	82	X
Lebanon	1,315	444	1,002	224.9	28	X	4	X	68	X
Malaysia	21,259	768	3,310	X	23	X	30	X	47	X
Mongolia	9,375	271	8,765	X	11	X	27	X	62	X
Myanmar (Burma)	22,719	101	3,276	X	7	X	3	X	90	X
Nepal	7,338	154	X	X	4	X	1	X	95	X
Oman	393	656	381	280.7	5	X	2	X	94	X
Pakistan	1,678	1,269	372	527.6	2	X	2	11.1	97	88.9
Philippines	4,476	686	2,494	82.8	18	0.0	21	50.0	61	50.0
Saudi Arabia	119	1,003	109	587.4	9	5.4	1	8.1	90	86.5
Singapore	172	84	X	X	45	X	51	X	4	X
Sri Lanka	2,341	503	921	X	2	X	2	X	96	X
Syria	456	1,069	280	353.0	4	X	2	X	94	X
Tajikistan	11,171	2,001	1,011	X	3	X	4	X	92	X
Thailand	1,845	602	721	15.0	4	60.0	6	25.7	90	14.3
Turkey	3,074	544	314	112.3	16	42.9	11	0.0	72	57.1
Turkmenistan	232	5,723	778	X	1	X	1	X	98	X
United Arab Emirates	64	954	51	251.3	7	X	1	X	62	X
Uzbekistan	704	2,501	848	X	4	X	5	X	94	X
Vietnam	4,827	416	1,078	X	13	X	9	X	78	X
Yemen	243	253	89	139.2	7	X	1	X	92	X

(Continued on next page)

| COUNTRY | Annual Renewable Water Resources[a] | | Annual Average Groundwater Resources[b] | | Sectoral Withdrawals (%)[c] | | | | | |
| | Supply Per Capita (cubic meters) 1998 est. | Withdrawal Per Capita (cubic meters) 1998 | Recharge Per Capita (cubic meters) 1998 est. | Withdrawal Per Capita (cubic meters) 1998 | Domestic | | Industry | | Agriculture | |
					Surface	Ground	Surface	Ground	Surface	Ground
EUROPE	**8,547**	625	X	X	14	X	55	X	31	X
Albania	2,903	94	2,032	X	6	X	18	X	76	X
Austria	6,857	304	2,716	144.7	33	52.1	58	42.7	9	5.1
Belarus	3,595	264	1,740	106.0	22	55.6	43	14.1	35	30.3
Belgium	822	917	84	79.2	11	68.3	85	27.0	4	4.8
Bosnia-Herzegovina	X	X	X	X	X	X	X	X	X	X
Bulgaria	2,146	1,574	1,598	566.0	3	X	76	X	22	X
Croatia	13,663	X	X	X	X	X	X	X	X	X
Czech Republic	5,694	266	X	77.9	41	X	57	X	2	X
Denmark	2,092	233	5,706	215.1	30	40.2	27	22.0	43	37.9
Estonia	8,642	107	2,718	X	56	X	39	X	5	X
Finland	21,334	440	369	47.3	12	64.9	85	10.8	3	24.3
France	3,065	665	1,703	109.5	16	52.5	69	30.2	15	17.3
Germany	1,165	580	555	97.4	11	48.6	70	47.5	20	3.9
Greece	4,279	523	237	193.9	8	12.8	29	2.7	63	84.5
Hungary	604	660	685	99.1	9	35.0	55	47.5	36	17.5
Iceland	606,498	636	86,643	394.2	31	X	63	X	6	X
Ireland	13,187	233	971	50.0	16	34.5	74	36.8	10	28.7
Italy	2,785	986	524	211.4	14	53.1	27	13.3	59	33.7
Latvia	6,685	114	879	X	55	X	32	X	13	X
Lithuania	4,174	68	322	X	81	X	16	X	3	X
Macedonia	X	X	X	X	X	X	X	X	X	X
Moldova	225	667	90	X	9	X	65	X	26	X
Netherlands	635	518	286	78.7	5	32.0	61	44.5	34	23.4
Norway	87,691	488	21,923	26.5	20	27.3	72	72.7	8	0.0
Poland	1,278	321	931	63.2	13	70.0	76	30.0	11	0.0
Portugal	3,878	738	521	210.6	15	38.6	37	22.8	48	38.6
Romania	1,639	1,139	368	55.5	8	61.0	33	38.1	59	0.8
Russian Federation	29,115	521	5,320	X	19	X	62	X	20	X
Slovak Republic	5,745	337	X	142.5	X	X	X	X	X	X
Slovenia	X	X	X	X	X	X	X	X	X	X
Spain	2,775	781	521	140.0	12	X	26	22.2	62	77.8
Sweden	19,858	340	2,257	69.8	36	91.7	55	8.3	9	0.0
Switzerland	5,802	173	369	138.6	23	94.7	73	5.3	4	0.0
Ukraine	1,029	504	388	82.8	18	30.3	52	17.5	30	52.1
United Kingdom	1,219	204	168	47.1	20	51.3	77	46.6	3	2.1
Yugoslavia (Serbia-Montenegro)	X	X	X	X	X	X	X	X	X	X

(Continued on next page)

COUNTRY	Annual Renewable Water Resources[a]		Annual Average Groundwater Resources[b]		Sectoral Withdrawals (%)[c]					
	Supply Per Capita (cubic meters) 1998 est.	Withdrawal Per Capita (cubic meters) 1998	Recharge Per Capita (cubic meters) 1998 est.	Withdrawal Per Capita (cubic meters) 1998	Domestic		Industry		Agriculture	
					Surface	Ground	Surface	Ground	Surface	Ground
OCEANIA	**54,795**	**X**	**X**	**X**	**64**	**X**	**2**	**X**	**34**	**X**
Australia	18,596	933	X	162.2	65	X	2	22.6	33	77.4
Fiji	34,732	42	87,591	X	20	X	20	X	60	X
New Zealand	88,859	589	53,804	X	46	X	10	X	44	X
Papua New Guinea	174,055	28	X	X	29	X	22	X	49	X
Solomon Islands	107,194	0	X	X	40	X	20	X	40	X

[a] Annual renewable water resources usually include river flows from other countries.

[b] Withdrawal data from most recent year available; varies by country from 1987 to 1995.

[c] Total withdrawals may exceed 100% because of groundwater withdrawals or river inflows.

Source: World Resources 1998–99 (World Resources Institute).

Part IX

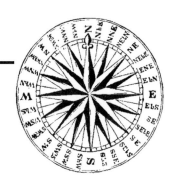

Geographic Index

The geographic index contains approximately 1500 names of cities, states, countries, rivers, lakes, mountain ranges, oceans, capes, bays, and other geographic features. The name of each geographical feature in the index is accompanied by a geographical coordinate (latitude and longitude) in degrees and by the page number of the primary map on which the geographical feature appears. Where the geographical coordinates are for specific places or *points,* such as a city or a mountain peak, then the latitude and longitude figures give the location of the map symbol denoting that point. Thus, Los Angeles, California, is at 34N and 118W and the location of Mt. Everest is 28N and 87E. The coordinates for political features (countries or states) or physical features (oceans, deserts) that are *areas* rather than *points* are given according to the location of the name of the feature on the map, except in those cases where the name of the feature is separated from the feature (a country's name appearing over an adjacent ocean area because of space requirements). In such cases, the feature's coordinates will indicate the location of the center of the feature. The coordinates for the Sahara Desert will lead the reader to the place name "Sahara Desert" on the map; the coordinates for North Carolina will show the center location of the state since the name appears over the adjacent Atlantic Ocean. Finally, the coordinates for geographical features that are *lines* rather than *points* or *areas* will also appear near the center of the text identifying the geographical feature. Alphabetizing follows general conventions; the names of physical features such as lakes, rivers, mountains are given as: proper name, followed by the generic name. Thus "Mount Everest" is listed as "Everest, Mt." Where an article such as "the," "le," or "al" appears in a geographic name, the name is alphabetized according to the article. Hence, "La Paz" is found under "L" and not under "P."

Name/Description	Latitude & Longitude	Page	Name/Description	Latitude & Longitude	Page
Abidjan, Cote d'Ivoire *(city, nat. cap.)*	5N 4W	99	Agulhas, Cape	35S 20E	100
Abu Dhabi, U.A.E. *(city, nat. cap.)*	24N 54E	103	Ahaggar Range	23N 6E	100
Accra, Ghana *(city, nat. cap.)*	64N 0	99	Ahmadabad, India *(city)*	23N 73E	103
Aconcagua, Mt. 22,881	38S 78W	91	Akmola, Kazakhstan *(city)*	51N 72E	103
Acre *(st., Brazil)*	9S 70W	90	Al Fashir, Sudan *(city)*	14N 25E	99
Addis Ababa, Ethiopia *(city, nat. cap.)*	9N 39E	99	Al Fayyum, Egypt *(city)*	29N 31E	99
Adelaide, S. Australia *(city, st. cap., Aust.)*	35S 139E	107	Al Hijaz Range	30N 40E	104
Aden, Gulf of	12N 46E	104	Al Khufra Oasis	24N 23E	100
Aden, Yemen *(city)*	13N 45E	103	Alabama *(st., US)*	33N 87W	86
Admiralty Islands	1S 146E	108	Alagoas *(st., Brazil)*	9S 37W	90
Adriatic Sea	44N 14E	95	Alaska *(st., US)*	63N 153W	86 inset
Aegean Sea	39N 25E	95	Alaska, Gulf of	58N 150W	86 inset
Afghanistan *(country)*	35N 65E	103	Alaska Peninsula	57N 155W	87 inset
Aguascalientes *(st., Mex.)*	22N 110W	86	Alaska Range	60N 150W	87 inset
Aguascalientes, Aguas. *(city, st. cap., Mex.)*	22N 102W	86	Albania *(country)*	41N 20E	94

Name/Description	Latitude & Longitude	Page
Belem, Para (city, st. cap., Braz.)	1S 48W	90
Belfast, Northern Ireland (city)	55N 6W	94
Belgium (country)	51N 4E	94
Belgrade, Yugoslavia (city, nat. cap.)	45N 21E	94
Belhuka, Mt. 14,483	50N 86E	104
Belize (country)	18S 88W	86
Belle Isle, Strait of	52N 57W	87
Belmopan, Belize (city, nat. cap.)	18S 89W	86
Belo Horizonte, M.G. (city, st. cap., Braz.)	20S 43W	90
Belyando (riv., Australasia)	22S 147W	108
Ben, Rio (riv., S.Am.)	14S 67W	91
Bengal, Bay of	15N 90E	104
Benguela, Angola (city)	13S 13E	99
Benin (country)	10N 4E	99
Benin City, Nigeria (city)	6N 6E	99
Benue (riv., Africa)	8N 9E	100
Bergen, Norway (city)	60N 5E	94
Bering Sea	57N 175W	104
Bering Strait	65N 168W	104
Berlin, Germany (city)	52N 13E	94
Bermeo, Rio (riv., S.Am.)	25S 61W	91
Bermuda (island)	30S 66W	86
Bhutan (country)	28N 110E	103
Billings, MT (city)	46N 108W	86
Birmingham, AL (city)	34N 87W	86
Birmingham, UK (city)	52N 2W	94
Biscay, Bay of	45N 5W	95
Bishkek, Kyrgyzstan (city, nat. cap.)	43N 75E	103
Bismarck Archipelago	4S 147E	108
Bismarck, North Dakota (city, st. cap., US)	47N 101W	86
Bismarck Range	6S 145E	108
Bissau, Guinea-Bissau (city, nat. cap.)	12N 16W	99
Black Sea	46N 34E	95
Blanc, Cape	21N 18W	100
Blue Nile (riv., Africa)	10N 36E	100
Blue Mountains	33S 150E	108
Boa Vista do Rio Branco, Roraima (city, st. cap., Braz.)	3N 61W	90
Boise, Idaho (city, st. cap., US)	44N 116W	86
Bolivia (country)	17S 65W	90
Boma, Congo Republic (city)	5S 13E	99
Bombay, India (city)	19N 73E	103
Bonn, Germany (city, nat. cap.)	51N 7E	94
Boothia Peninsula	71N 94W	87
Borneo (island)	0 11E	103
Bosnia and Herzegovina (country)	45N 18E	94
Bosporus, Strait of	41N 29E	95
Boston, Massachusetts (city, st. cap., US)	42N 71W	86
Botany Bay	35S 153E	107
Bothnia, Gulf of	62N 20E	95
Botswana (country)	23S 25E	99
Brahmaputra (riv., Asia)	30N 100E	104
Branco, Rio (riv., S.Am.)	3N 62W	91
Brasilia, Brazil (city, nat. cap.)	16S 48W	90
Bratislava, Slovakia (city, nat. cap.)	48N 17E	94
Brazil (country)	10S 52W	90
Brazilian Highlands	18S 45W	91
Brazzaville, Congo (city, nat. cap.)	4S 15E	99
Brisbane, Queensland (city, st. cap., Aust.)	27S 153E	107
Bristol Bay	58N 159W	87 inset
British Columbia (prov., Can.)	54N 130W	86
Brooks Range	67N 155W	87
Bruce, Mt. 4052	22S 117W	108
Brussels, Belgium (city, nat. cap.)	51N 4E	94
Bucharest, Romania (city, nat. cap.)	44N 26E	94
Budapest, Hungary (city, nat. cap.)	47N 19E	94
Buenos Aires, Argentina (city, nat. cap.)	34S 58W	90
Buenos Aires (st., Argentina)	36S 60W	90
Buffalo, NY (city)	43N 79W	86
Bujumbura, Burundi (city, nat. cap.)	3S 29E	99
Bulgaria (country)	44N 26E	94
Bur Sudan, Sudan (city)	19N 37E	99
Burdekin (riv., Australasia)	19S 146W	108
Burkina Faso (country)	11N 2W	99
Buru (island)	4S 127E	108
Burundi (country)	4S 30E	99
Cairns, Aust. (city)	17S 145E	107
Cairo, Egypt (city, nat. cap.)	30N 31E	99
Calcutta, India (city)	23N 88E	103
Calgary, Canada (city)	51N 114W	86
Calicut, India (city)	11N 76E	103
California (st., US)	35N 120W	86
California, Gulf of	29N 110W	86
Callao, Peru (city)	13S 77W	90
Cambodia (country)	10N 106E	103
Cameroon (country)	5N 13E	99
Campeche (st., Mex.)	19N 90W	86
Campeche Bay	20N 92W	87
Campeche, Campeche (city, st. cap., Mex.)	19N 90W	86
Campo Grande, M.G.S. (city, st. cap., Braz.)	20S 55W	90
Canada (country)	52N 100W	86
Canadian (riv., N.Am.)	30N 100W	87
Canary Islands	29N 18W	100
Canberra, Australia (city, nat. cap.)	35S 149E	107
Cape Breton Island	46N 60W	87
Cape Town, South Africa (city)	34S 18E	99
Caracas, Venezuela (city, nat. cap.)	10N 67W	90
Caribbean Sea	18N 75W	87
Carnarvon, Australia (city)	25S 113E	107
Carpathian Mountains	48N 24E	95
Carpentaria, Gulf of	14S 140E	108
Carson City, Nevada (city, st. cap., US)	39N 120W	86
Cartagena, Colombia (city)	10N 76W	90
Cascade Range	45N 120W	87
Casiquiare, Rio (riv., S.Am.)	4N 67W	91
Caspian Depression	49N 48E	95
Caspian Sea	42N 48E	95
Catamarca (st., Argentina)	25S 70W	90
Catamarca, Catamarca (city, st. cap., Argen.)	28S 66W	90
Cauca, Rio (riv., S.Am.)	8N 75W	91
Caucasus Mountains	42N 40E	95
Cayenne, French Guiana (city, nat. cap.)	5N 52W	90
Ceara (st., Brazil)	4S 40W	90
Celebes (island)	0 120E	104
Celebes Sea	2N 120E	104
Central African Republic (country)	5N 20E	99

Name/Description	Latitude & Longitude	Page	Name/Description	Latitude & Longitude	Page
Ceram (island)	3S 129E	107	Connecticut (st., US)	43N 76W	86
Chaco (st., Argentina)	25S 60W	90	Connecticut (riv., N.Am.)	43N 76W	87
Chad (country)	15N 20E	99	Cook, Mt. 12,316	44S 170E	108
Chad, Lake	12N 12E	99	Cook Strait	42S 175E	108
Changchun, China (city)	44N 125E	103	Copenhagen, Denmark (city, nat. cap.)	56N 12E	94
Chari (riv., Africa)	11N 16E	100	Copiapo, Chile (city)	27S 70W	90
Charleston, SC (city)	33N 80W	86	Copiapo , Mt. 19,947	26S 70W	91
Charleston, West Virginia (city, st. cap., US)	38N 82W	86	Coquimbo, Chile (city)	30S 70W	90
Charlotte, NC (city)	35N 81W	86	Coral Sea	15S 155E	108
Charlotte Waters, Aust. (city)	26S 135E	107	Cordilleran Highlands	45N 118W	87
Charlottetown, P.E.I. (city, prov. cap., Can.)	46N 63W	86	Cordoba (st., Argentina)	32S 67W	90
Chelyabinsk, Russia (city)	55N 61E	94	Cordoba, Cordoba (city, st. cap., Argen.)	32S 64W	90
Chengdu, China (city)	30N 104E	103	Corrientes (st., Argentina)	27S 60W	90
Chesapeake Bay	36N 74W	87	Corrientes, Corrientes (city, st. cap., Argen.)	27S 59W	90
Chetumal, Quintana Roo (city, st. cap., Mex.)	19N 88W	86	Corsica (island)	42N 9E	94
Cheyenne, Wyoming (city, st. cap., US)	41N 105W	86	Cosmoledo Islands	9S 48E	100
Chiapas (st., Mex.)	17N 92W	86	Costa Rica (country)	15N 84W	86
Chicago, IL (city)	42N 87W	86	Cote d'Ivoire (country)	7N 86W	99
Chiclayo, Peru (city)	7S 80W	90	Cotopaxi, Mt. 19,347	1S 78W	91
Chidley, Cape	60N 65W	87	Crete (island)	36N 25W	95
Chihuahua (st., Mex.)	30N 110W	86	Croatia (country)	46N 20W	94
Chihuahua, Chihuahua (city, st. cap., Mex.)	29N 106W	86	Cuango (riv., Africa)	10S 16E	100
Chile (country)	32S 75W	90	Cuba (country)	22N 78W	86
Chiloe (island)	43S 74W	91	Cuiaba, Mato Grosso (city, st. cap., Braz.)	16S 56W	90
Chilpancingo, Guerrero (city, st. cap., Mex.)	19N 99W	86	Ciudad Victoria, Tamaulipas (city, st. cap., Mex.)	24N 99W	86
Chimborazo, Mt. 20,702	2S 79W	91	Culiacan, Sinaloa (city, st. cap., Mex.)	25N 107W	86
China (country)	38N 105E	103	Curitiba, Parana (city, st. cap., Braz.)	26S 49W	90
Chisinau, Moldova (city, nat. cap.)	47N 29E	94	Cusco, Peru (city)	14S 72W	90
Chongqing, China (city)	30N 107E	103	Cyprus (island)	36N 34E	95
Christchurch, New Zealand (city)	43S 173E	107	Czech Republic (country)	50N 16E	94
Chubut (st., Argentina)	44S 70W	90	d'Ambre, Cape	12S 50E	99
Chubut, Rio (riv., S.Am.)	44S 71W	91	Dakar, Senegal (city, nat. cap.)	15N 17W	99
Cincinnati, OH (city)	39N 84W	86	Dakhla, Western Sahara (city)	24N 16W	99
Cleveland (city)	41N 82W	86	Dallas, TX (city)	33N 97W	86
Coahuila (st., Mex.)	30N 105W	86	Dalrymple , Mt. 4190	22S 148E	108
Coast Mountains (Can.)	55N 130W	87	Daly (riv., Australasia)	14S 132E	108
Coast Ranges (US)	40N 120W	87	Damascus, Syria (city, nat. cap.)	34N 36E	94
Coco Island	8N 88W	87	Danube (riv., Europe)	44N 24E	95
Cod, Cape	42N 70W	87	Dar es Salaam, Tanzania (city, nat. cap.)	7S 39E	99
Colima (st., Mex.)	18N 104W	86	Darien, Gulf of	9N 77W	91
Colima, Colima (city, st. cap., Mex.)	19N 104W	86	Darling (riv., Australasia)	35S 144E	108
Colombia (country)	4N 73W	90	Darling Range	33S 116W	108
Colombo, Sri Lanka (city, nat. cap.)	7N 80E	103	Darwin, Northern Terr. (city, st. cap., Aust.)	12S 131E	107
Colorado (riv., N.Am.)	36N 110W	87	Davis Strait	57N 59W	87
Colorado (st., US)	38N 104W	86	Deccan Plateau	20N 80E	104
Colorado, Rio (riv., S.Am.)	38S 70W	91	DeGrey (riv., Australasia)	22S 120E	108
Colorado (Texas) (riv., N.Am.)	30N 100W	87	Delaware (st., US)	38N 75W	86
Columbia (riv., N.Am.)	45N 120W	86	Delaware (riv., N.Am.)	38N 77W	87
Columbia, South Carolina (city, st. cap., US)	34N 81W	86	Delhi, India (city)	30N 78E	103
Columbus, Ohio (city, st. cap., US)	40N 83W	86	Denmark (country)	55N 10E	94
Comodoro Rivadavia, Argentina (city)	68S 70W	90	Denmark Strait	67N 27W	95
Comoros (country)	12S 44E	99	D'Entrecasteaux Islands	10S 153E	108
Conakry, Guinea (city, nat. cap.)	9N 14W	94	Denver, Colorado (city, st. cap., US)	40N 105W	86
Concord, New Hampshire (city, st. cap., US)	43N 71W	86	Derby, Australia (city)	17S 124E	107
Congo (country)	3S 15E	99	Des Moines (riv., N.Am.)	43N 95W	87
Congo (riv., Africa)	3N 22E	100	Des Moines, Iowa (city, st. cap., US)	42N 92W	86
Congo Basin	4N 22E	100	Desolacion Island	54S 73W	91
Congo, Democratic Republic of (country)	5S 15E	99	Detroit, MI (city)	42N 83W	86

Name/Description	Latitude & Longitude	Page	Name/Description	Latitude & Longitude	Page
Dhaka, Bangladesh (city, nat. cap.)	24N 90E	103	Falkland Islands (Islas Malvinas)	52S 60W	91
Dinaric Alps	44N 20E	95	Farewell, Cape (NZ)	40S 170E	108
Djibouti (country)	12N 43E	99	Fargo, ND (city)	47N 97W	86
Djibouti, Djibouti (city, nat. cap.)	12N 43E	99	Farquhar, Cape	24S 114E	108
Dnepr (riv., Europe)	50N 34E	95	Fiji (country)	17S 178E	107
Dnipropetrovsk, Ukraine (city)	48N 35E	94	Finisterre, Cape	44N 10W	95
Dodoma, Tanzania (city)	6S 36E	99	Finland (country)	62N 28E	94
Dominican Republic (country)	20N 70W	86	Finland, Gulf of	60N 20E	95
Don (riv., Europe)	53N 39E	95	Firth of Forth	56N 3W	95
Donetsk, Ukraine (city)	48N 38E	94	Fitzroy (riv., Australasia)	17S 125E	108
Dover, Delaware (city, st. cap., US)	39N 75W	86	Flinders Range	31S 139E	108
Dover, Strait of	52N 0	95	Flores (island)	8S 121E	108
Drakensberg	30S 30E	100	Florianopolis, Sta. Catarina (city, st. cap., Braz.)	27S 48W	90
Dublin, Ireland (city, nat. cap.)	53N 6W	94	Florida (st., US)	28N 83W	86
Duluth, MN (city)	47N 92W	86	Florida, Strait of	28N 80W	87
Dunedin, New Zealand (city)	46S 171E	107	Fly (riv., Australasia)	8S 143E	108
Durango (st., Mex.)	25N 108W	86	Formosa (st., Argentina)	23S 60W	90
Durango, Durango (city, st. cap., Mex.)	24N 105W	86	Formosa, Formosa (city, st. cap., Argen.)	27S 58W	90
Durban, South Africa (city)	30S 31E	99	Fort Worth, TX (city)	33N 97W	86
Dushanbe, Tajikistan (city, nat. cap.)	39N 69E	103	Fortaleza, Ceara (city, st. cap., Braz.)	4S 39W	90
Dvina (riv., Europe)	64N 42E	95	France (country)	46N 4E	94
Dzhugdzhur Khrebet	58N 138E	104	Frankfort, Kentucky (city, st. cap., US)	38N 85W	86
East Cape (NZ)	37S 180E	108	Frankfurt, Germany (city)	50N 9E	94
East China Sea	30N 128E	104	Fraser (riv., N.Am.)	52N 122W	87
Eastern Ghats	15N 80E	104	Fredericton, N.B. (city, prov. cap., Can.)	46N 67W	86
Ecuador (country)	3S 78W	90	Fremantle, Australia (city)	33S 116E	107
Edmonton, Alberta (city, prov. cap., Can.)	54N 114W	86	Freetown, Sierra Leone (city, nat. cap.)	8N 13W	99
Edward, Lake	0 30E	100	French Guiana (country)	4N 52W	90
Egypt (country)	23N 30E	99	Fria, Cape	18S 12E	100
El Aaiun, Western Sahara (city)	27N 13W	99	Fuzhou, China (city)	26N 119E	103
El Djouf	25N 15W	100	Gabes, Gulf of	33N 12E	100
El Paso, TX (city)	32N 106W	86	Gabes, Tunisia (city)	34N 10E	99
El Salvador (country)	15N 90W	86	Gabon (country)	2S 12E	99
Elbe (riv., Europe)	54N 10E	95	Gaborone, Botswana (city, nat. cap.)	25S 25E	99
Elburz Mountains	28N 60E	104	Gairdiner, Lake	32S 136E	108
Elbruz, Mt. 18,510	43N 42E	104	Galveston, TX (city)	29N 95W	86
Elgon, Mt. 14,178	1N 34E	100	Gambia (country)	13N 15W	99
English Channel	50N 0	95	Gambia (riv., Africa)	13N 15W	100
Entre Rios (st., Argentina)	32S 60W	90	Ganges (riv., Asia)	27N 85E	104
Equatorial Guinea (country)	3N 10E	99	Gascoyne (riv., Australasia)	25S 115E	108
Erg Iguidi	26N 6W	100	Gaspé Peninsula	50N 70W	87
Erie (lake, N.Am.)	42N 85W	87	Gdansk, Poland (city)	54N 19E	94
Eritrea (country)	16N 38E	99	Geelong, Aust. (city)	38S 144E	107
Erzegebirge Mountains	50N 14E	95	Gees Gwardafuy (island)	15N 50E	100
Espinhaco Mountains	15S 42W	91	Genoa, Gulf of	44N 10E	95
Espiritu Santo (island)	15S 168E	107	Geographe Bay	35S 115E	108
Espiritu Santo (st., Brazil)	20S 42W	90	Georgetown, Guyana (city, nat. cap.)	8N 58W	90
Essen, Germany (city)	52N 8E	94	Georgia (country)	42N 44E	94
Estonia (country)	60N 26E	94	Georgia (st., US)	30N 82W	86
Ethiopia (country)	8N 40E	99	Germany (country)	50N 12E	94
Ethiopian Plateau	8N 40E	100	Ghana (country)	8N 3W	99
Euphrates (riv., Asia)	28N 50E	104	Gibraltar, Strait of	37N 6W	95
Everard, Lake	32S 135E	108	Gibson Desert	24S 124E	108
Everard Ranges	28S 135E	108	Gilbert (riv., Australasia)	8S 142E	108
Everest, Mt. 29,028	28N 84E	104	Giluwe, Mt. 14,330	5S 144E	108
Eyre, Lake	29S 136E	108	Glasgow, Scotland (city)	56N 6W	94
Faeroe Islands	62N 11W	95	Gobi Desert	48N 105E	104
Fairbanks, AK (city)	63N 146W	86	Godavari (riv., Asia)	18N 82E	104

Name/Description	Latitude & Longitude	Page	Name/Description	Latitude & Longitude	Page
Godwin-Austen (K2), Mt. 28,250	30N 70E	104	Harbin, China *(city)*	46N 126E	103
Goiania, Goias *(city, st. cap., Braz.)*	17S 49W	90	Harer, Ethiopia *(city)*	10N 42E	99
Goias *(st., Brazil)*	15S 50W	90	Hargeysa, Somalia *(city)*	9N 44E	99
Gongga Shan 24,790	26N 102E	104	Harrisburg, Pennsylvania *(city, st. cap., US)*	40N 77W	86
Good Hope, Cape of	33S 18E	100	Hartford, Connecticut *(city, st. cap., US)*	42N 73W	86
Goteborg, Sweden *(city)*	58N 12E	94	Hatteras, Cape	32N 73W	87
Gotland *(island)*	57N 20E	95	Havana, Cuba *(city, nat. cap.)*	23N 82W	86
Grampian Mountains	57N 4W	95	Hawaii *(st., US)*	21N 156W	87 inset
Gran Chaco	23S 70N	91	Hebrides *(island)*	58N 8W	95
Grand Erg Occidental	29N 0	100	Helena, Montana *(city, st. cap., US)*	47N 112W	86
Grand Teton 13, 770	45N 112W	87	Helsinki, Finland *(city, nat. cap.)*	60N 25E	94
Great Artesian Basin	25S 145E	108	Herat, Afghanistan *(city)*	34N 62E	103
Great Australian Bight	33S 130E	108	Hermosillo, Sonora *(city, st. cap., Mex.)*	29N 111W	86
Great Barrier Reef	15S 145E	108	Hidalgo *(st., Mex.)*	20N 98W	86
Great Basin	39N 117W	87	Himalayas	26N 80E	104
Great Bear Lake *(lake, N.Am.)*	67N 120W	87	Hindu Kush	30N 70E	104
Great Dividing Range	20S 145E	108	Ho Chi Minh City, Vietnam *(city)*	11N 107E	103
Great Indian Desert	25N 72E	104	Hobart, Tasmania *(city, st. cap., Aust.)*	43S 147E	107
Great Namaland	25S 16E	100	Hokkaido *(island)*	43N 142E	104
Great Plains	40N 105W	87	Honduras *(country)*	16N 87W	86
Great Salt Lake *(lake, N.Am.)*	40N 113W	87	Honduras, Gulf of	15N 88W	87
Great Sandy Desert	23S 125E	108	Honiara, Solomon Islands *(city, nat. cap.)*	9S 160E	107
Great Slave Lake *(lake, N.Am.)*	62N 110W	87	Honolulu, Hawaii *(city, st. cap., US)*	21N 158W	86 inset
Great Victoria Desert	30S 125E	108	Honshu *(island)*	38N 140E	104
Greater Khingan Range	50N 120E	104	Hormuz, Strait of	25N 58E	104
Greece *(country)*	39N 21E	94	Horn, Cape	55S 70W	91
Greenland (Denmark) *(country)*	78N 40W	86	Houston, TX *(city)*	30N 95W	86
Gregory Range	18S 145E	108	Howe, Cape	37S 150E	108
Grey Range	26S 145E	108	Huambo, Angola *(city)*	13S 16E	99
Guadalajara, Jalisco *(city, st. cap., Mex.)*	21N 103W	86	Huang *(riv., Asia)*	30N 105E	104
Guadalcanal *(island)*	9S 160E	107	Huascaran, Mt. 22,133	8N 79W	91
Guadeloupe *(island)*	29N 120W	87	Hudson *(riv., N.Am.)*	42N 76W	86
Guanajuato *(st., Mex.)*	22N 100W	86	Hudson Bay	60N 90W	87
Guanajuato, Guanajuato *(city, st. cap., Mex.)*	21N 101W	86	Hudson Strait	63N 70W	87
Guangzhou, China *(city)*	23N 113E	103	Hue, Vietnam *(city)*	15N 110E	103
Guapore, Rio *(riv., S.Am.)*	15S 63W	91	Hughes, Aust. *(city)*	30S 130E	107
Guatemala *(country)*	14N 90W	86	Hungary *(country)*	48N 20E	94
Guatemala, Guatemala *(city, nat. cap.)*	15N 91W	86	Huron *(lake, N.Am.)*	45N 85W	87
Guayaquil, Ecuador *(city)*	2S 80W	90	Hyderabad, India *(city)*	17N 79E	103
Guayaquil, Gulf of	3S 83W	91	Ibadan, Nigeria *(city)*	7N 4E	99
Guerrero *(st., Mex.)*	18N 102W	86	Iceland *(country)*	64N 20W	94
Guianas Highlands	5N 60W	91	Idaho *(st., US)*	43N 113W	86
Guinea *(country)*	10N 10W	99	Iguassu Falls	25S 55W	91
Guinea, Gulf of	3N 0	100	Illimani, Mt. 20,741	16S 67W	91
Guinea-Bissau *(country)*	12N 15W	99	Illinois *(riv., N.Am.)*	40N 90W	87
Guyana *(country)*	6N 57W	90	Illinois *(st., US)*	44N 90W	86
Gydan Range	62N 155E	104	India *(country)*	23N 80E	103
Haiti *(country)*	18N 72W	86	Indiana *(st., US)*	46N 88W	86
Hakodate, Japan *(city)*	42N 140E	103	Indianapolis, Indiana *(city, st. cap., US)*	40N 86W	86
Halifax Bay	18S 146E	108	Indigirka *(riv., Asia)*	70N 145E	104
Halifax, Nova Scotia *(city, prov. cap., Can.)*	45N 64W	86	Indonesia *(country)*	2S 120E	103
Halmahera *(island)*	1N 128E	104 inset	Indus *(riv., Asia)*	25N 70E	104
Hamburg, Germany *(city)*	54N 10E	94	Ionian Sea	38N 19E	95
Hammersley Range	23S 116W	108	Iowa *(st., US)*	43N 95W	86
Hann, Mt. 2,800	15S 127E	108	Iquitos, Peru *(city)*	4S 74W	90
Hanoi, Vietnam *(city, nat. cap.)*	21N 106E	103	Iran *(country)*	30N 55E	103
Hanover Island	52S 74W	91	Iraq *(country)*	30N 50E	103
Harare, Zimbabwe *(city, nat. cap.)*	18S 31E	99	Ireland *(country)*	54N 8W	94

Name/Description	Latitude & Longitude	Page	Name/Description	Latitude & Longitude	Page
Latvia *(country)*	56N 24E	94	Malacca, Strait of	3N 100E	103
Laurentian Highlands	48N 72W	87	Malawi *(country)*	13S 35E	99
Lebanon *(country)*	34N 35E	94	Malaysia *(country)*	3N 110E	103
Leeds, UK *(city)*	54N 2W	94	Malekula *(island)*	16S 166E	108
Le Havre, France *(city)*	50N 0	94	Mali *(country)*	17N 5W	99
Lena *(riv., Asia)*	70N 125E	104	Malpelo Island	8N 84W	87
Lesotho *(country)*	30S 27E	99	Malta *(island)*	36N 16E	95
Leveque, Cape	16S 123E	108	Mamore, Rio *(riv., S.Am.)*	15S 65W	91
Leyte *(island)*	12N 130E	104	Managua, Nicaragua *(city, nat. cap.)*	12N 86W	86
Lhasa, Tibet (China) *(city)*	30N 91E	103	Manaus, Amazonas *(city, st. cap., Braz.)*	3S 60W	90
Liberia *(country)*	6N 10W	99	Manchester, UK *(city)*	53N 2W	94
Libreville, Gabon *(city, nat. cap.)*	0 9E	99	Mandalay, Myamar *(city)*	22N 96E	103
Libya *(country)*	27N 17E	99	Manila, Philippines *(city, nat. cap.)*	115N 121E	103
Libyan Desert	27N 25E	100	Manitoba *(prov., Can.)*	52N 93W	86
Lille, France *(city)*	51N 3E	94	Mannar, Gulf of	9N 79E	104
Lilongwe, Malawi *(city, nat. cap.)*	14S 33E	99	Maoke Mountains	5S 138E	108
Lima, Peru *(city, nat. cap.)*	12S 77W	90	Maputo, Mozambique *(city, nat. cap.)*	26S 33E	99
Limpopo *(riv., Africa)*	22S 30E	100	Maracaibo, Lake	10N 72W	90
Lincoln, Nebraska *(city, st. cap., US)*	41N 97W	86	Maracaibo, Venezuela *(city)*	11N 72W	90
Lisbon, Portugal *(city, nat. cap.)*	39N 9W	94	Maracapa, Amapa *(city, st. cap., Braz.)*	0 51W	90
Lithuania *(country)*	56N 24E	94	Maranhao *(st., Brazil)*	4S 45W	90
Little Rock, Arkansas *(city, st. cap., US)*	35N 92W	86	Maranon, Rio *(riv., S.Am.)*	5S 75W	91
Liverpool, UK *(city)*	53N 3W	94	Marseille, France *(city)*	43N 5E	94
Ljubljana, Slovenia *(city, nat. cap.)*	46N 14E	94	Maryland *(st., US)*	37N 76W	86
Llanos	33N 103W	91	Masai Steppe	5S 35E	100
Logan, Mt. 18,551	62N 139W	87	Maseru, Lesotho *(city, nat. cap.)*	29S 27E	99
Logone *(riv., Africa)*	10N 14E	100	Mashad, Iran *(city)*	36N 59E	103
Lome, Togo *(city, nat. cap.)*	6N 1E	99	Massachusetts *(st., US)*	42N 70W	86
London, United Kingdom *(city, nat. cap.)*	51N 0	94	Massif Central	45N 3E	95
Londonderry, Cape	14S 125E	108	Mato Grosso	16S 52W	91
Lopez, Cape	1S 8E	100	Mato Grosso *(st., Brazil)*	15S 55W	90
Los Angeles, CA *(city)*	34N 118W	86	Mato Grosso do Sul *(st., Brazil)*	20S 55W	90
Los Chonos Archipelago	45S 74W	91	Mauritania *(country)*	20N 10W	99
Louisiana *(st., US)*	30N 90W	86	Mbandaka, Congo Republic *(city)*	0 18E	99
Lower Hutt, New Zealand *(city)*	45S 175E	107	McKinley, Mt. 20,320	62N 150W	87 inset
Luanda, Angola *(city, nat. cap.)*	9S 13E	99	Medellin, Colombia *(city)*	6N 76W	90
Lubumbashi, Congo Republic *(city)*	12S 28E	99	Mediterranean Sea	36N 16E	95
Lusaka, Zambia *(city, nat. cap.)*	15S 28E	99	Mekong *(riv., Asia)*	15N 108E	104
Luxembourg *(country)*	50N 6E	94	Melbourne, Victoria *(city, st. cap., Aust.)*	38S 145E	107
Luxembourg, Luxembourg *(city, nat. cap.)*	50N 6E	94	Melville, Cape	15S 145E	108
Luzon *(island)*	17N 121E	104	Memphis, TN *(city)*	35N 90W	86
Luzon Strait	20N 121E	104	Mendoza *(st., Argentina)*	35S 70W	90
Lyon, France *(city)*	46N 5E	94	Mendoza, Mendoza *(city, st. cap., Argen.)*	33S 69W	90
Lyon, Gulf of	42N 4E	95	Merida, Yucatan *(city, st. cap. Mex.)*	21N 90W	86
Maccio, Alagoas *(city, st. cap., Braz.)*	10S 36W	90	Merauke, New Guinea (Indon.) *(city)*	9S 140E	107
Macdonnell Ranges	23S 135E	108	Mexicali, Baja California *(city, st. cap., Mex.)*	32N 115W	86
Macedonia *(country)*	41N 21E	94	Mexico *(country)*	30N 110W	86
Mackenzie *(riv., N.Am.)*	68N 130W	87	Mexico *(st., Mex.)*	18N 100W	86
Macquarie *(riv., Australasia)*	33S 146E	108	Mexico City, Mexico *(city, nat. cap.)*	19N 99W	86
Madagascar *(country)*	20S 46E	99	Mexico, Gulf of	26N 90W	87
Madeira, Rio *(riv., S.Am.)*	5S 60W	91	Miami, FL *(city)*	26N 80W	86
Madison, Wisconsin *(city, st. cap., US)*	43N 89W	86	Michigan *(st., US)*	45N 82W	86
Madras, India *(city)*	13N 80E	103	Michigan *(lake, N.Am.)*	45N 90W	87
Madrid, Spain *(city, nat. cap.)*	40N 4W	94	Michoacan *(st., Mex.)*	17N 107W	86
Magdalena, Rio *(riv., S.Am.)*	8N 74W	91	Milan, Italy *(city)*	45N 9E	94
Magellan, Strait of	54S 68W	91	Milwaukee, WI *(city)*	43N 88W	86
Maine *(st., US)*	46N 70W	86	Minas Gerais *(st., Brazil)*	17S 45W	90
Malabo, Equatorial Guinea *(city, nat. cap.)*	4N 9E	99	Mindoro *(island)*	13N 120E	103

Name/Description	Latitude & Longitude	Page	Name/Description	Latitude & Longitude	Page
Minneapolis, MN *(city)*	45N 93W	86	Negros *(island)*	10N 125E	104
Minnesota *(st., US)*	45N 90W	86	Nelson *(riv., N.Am.)*	56N 90W	87
Minsk, Belarus *(city, nat. cap.)*	54N 28E	94	Nepal *(country)*	29N 85E	103
Misiones *(st., Argentina)*	25S 55W	90	Netherlands *(country)*	54N 6E	94
Mississippi *(riv., N.Am.)*	28N 90W	87	Neuquen *(st., Argentina)*	38S 68W	90
Mississippi *(st., US)*	30N 90W	86	Neuquen, Neuquen *(city, st. cap., Argen.)*	39S 68W	90
Missouri *(riv., N.Am.)*	41N 96W	87	Nevada *(st., US)*	37N 117W	86
Missouri *(st., US)*	35N 92W	86	New Britain *(island)*	5S 152E	108
Misti, Mt. 19,101	15S 73W	91	New Brunswick *(prov., Can.)*	47N 67W	86
Mitchell *(riv., Australasia)*	16S 143E	108	New Caledonia *(island)*	21S 165E	108
Mobile, AL *(city)*	31N 88W	86	New Delhi, India *(city, nat. cap.)*	29N 77E	103
Moçambique, Mozambique *(city)*	15S 40E	99	New Georgia *(island)*	8S 157E	108
Mogadishu, Somalia *(city, nat. cap.)*	2N 45E	99	New Guinea *(island)*	5S 142E	108
Moldova *(country)*	49N 28E	94	New Hampshire *(st., US)*	45N 70W	86
Mombasa, Kenya *(city)*	4S 40E	99	New Hanover *(island)*	3S 153E	108
Monaco, Monaco *(city)*	44N 8E	94	New Hebrides *(island)*	15S 165E	108
Mongolia *(country)*	45N 100E	103	New Ireland *(island)*	4S 154E	108
Monrovia, Liberia *(city, nat. cap.)*	6N 11W	99	New Jersey *(st., US)*	40N 75W	86
Montana *(st., US)*	50N 110W	86	New Mexico *(st., US)*	30N 108W	86
Monterrey, Nuevo Leon *(city, st. cap., Mex.)*	26N 100W	86	New Orleans, LA *(city)*	30N 90W	86
Montevideo, Uruguay *(city, nat. cap.)*	35S 56W	90	New Siberian Islands	74N 140E	104
Montgomery, Alabama *(city, st. cap., US)*	32N 86W	86	New South Wales *(st., Aust.)*	35S 145E	107
Montpelier, Vermont *(city, st. cap., US)*	44N 73W	86	New York *(city)*	41N 74W	86
Montreal, Canada *(city)*	45N 74W	86	New York *(st., US)*	45N 75W	86
Morelin, Michoacan *(city, st. cap., Mex.)*	20N 100W	86	New Zealand *(country)*	40S 170E	107
Morocco *(country)*	34N 10W	99	Newcastle, Aust. *(city)*	33S 152E	107
Moroni, Comoros *(city, nat. cap.)*	12S 42E	99	Newcastle, UK *(city)*	55N 2W	94
Moscow, Russia *(city, nat. cap.)*	56N 38E	94	Newfoundland *(prov., Can.)*	53N 60W	86
Mountain Nile *(riv., Africa)*	5N 30E	100	Nicaragua *(country)*	10N 90W	86
Mozambique *(country)*	19N 35E	99	Niamey, Niger *(city, nat. cap.)*	14N 2E	99
Mozambique Channel	19N 42E	100	Nicobar Islands	5N 93E	104
Munich, Germany *(city)*	48N 12E	94	Niger *(riv., Africa)*	12N 0	100
Murchison *(riv., Australasia)*	26S 115E	108	Niger *(country)*	10N 8E	99
Murmansk, Russia *(city)*	69N 33E	94	Nigeria *(country)*	8N 5E	99
Murray *(riv., Australasia)*	36S 143E	108	Nile *(riv., Africa)*	25N 31E	100
Murrumbidgee *(riv., Australasia)*	35S 146E	108	Nipigon *(lake, N.Am.)*	50N 87W	87
Muscat, Oman *(city, nat. cap.)*	23N 58E	103	Nizhny-Novgorod, Russia *(city)*	56N 44E	94
Musgrave Ranges	28S 135E	108	Norfolk, VA *(city)*	37N 76W	86
Myanmar (Burma) *(country)*	20N 95E	103	North Cape (NZ)	36N 174W	108
N. Saskatchewan *(riv., N.Am.)*	55N 110W	87	North Carolina *(st., US)*	30N 78W	86
Nairobi, Kenya *(city, nat. cap.)*	1S 37E	99	North Channel	56N 5W	95
Namibe, Angola *(city)*	16S 13E	99	North Dakota *(st., US)*	49N 100W	86
Namibia *(country)*	20S 16E	99	North Island (NZ)	37S 175W	108
Namoi *(riv., Australasia)*	31S 150E	108	North Sea	56N 3E	95
Nan Ling Mountains	25N 110E	104	North West Cape	22S 115W	108
Nanda Devi, Mt. 25,645	30N 80E	104	Northern Territory *(st., Aust.)*	20S 134W	108
Nanjing, China *(city)*	32N 119E	103	Northwest Territories *(prov., Can.)*	65N 125W	86
Nansei Shoto *(island)*	27N 125E	104	Norway *(country)*	62N 8E	94
Naples, Italy *(city)*	41N 14E	94	Nouakchott, Mauritania *(city, nat. cap.)*	18N 16W	99
Nashville, Tennessee *(city, st. cap., US)*	36N 87W	86	Noumea, New Caledonia *(city)*	22S 167E	107
Nasser, Lake	22N 32E	100	Nova Scotia *(prov., Can.)*	46N 67W	86
Natal, Rio Grande do Norte *(city, st. cap., Braz.)*	6S 5W	90	Novaya Zemlya *(island)*	72N 55E	104
Naturaliste, Cape	35S 115E	108	Novosibirsk, Russia *(city)*	55N 83E	103
Nayarit *(st., Mex.)*	22N 106W	86	Nubian Desert	20N 30E	100
N'Djamena, Chad *(city, nat. cap.)*	12N 15E	99	Nuevo Leon *(st., Mex.)*	25N 100W	86
Nebraska *(st., US)*	42N 100W	86	Nullarbor Plain	34S 125W	108
Negro, Rio (Argentina) *(riv., S.Am.)*	40S 70W	91	Nyasa, Lake	10S 35E	100
Negro, Rio (Brazil) *(riv., S.Am.)*	0 65W	91	Oakland, CA *(city)*	38N 122W	86

Name/Description	Latitude & Longitude	Page	Name/Description	Latitude & Longitude	Page
Oaxaca (st., Mex.)	17N 97W	86	Patagonia	43S 70W	91
Oaxaca, Oaxaca (city, st. cap., Mex.)	17N 97W	86	Paulo Afonso Falls	10S 40W	91
Ob (riv., Asia)	60N 78E	104	Peace (riv., N.Am.)	55N 120W	87
Ohio (riv., N.Am.)	38N 85W	87	Pennsylvania (st., US)	43N 80W	86
Ohio (st., US)	42N 85W	86	Pernambuco (st., Brazil)	7S 36W	90
Okavongo (riv., Africa)	18S 18E	100	Persian Gulf	28N 50E	104
Okavango Swamp	21S 23E	100	Perth, W. Australia (city, st. cap., Aust.)	32S 116E	107
Okeechobee (lake, N.Am.)	28N 82W	87	Peru (country)	10S 75W	90
Okhotsk, Russia (city)	59N 140E	103	Peshawar, Pakistan (city)	34N 72E	103
Okhotsk, Sea of	57N 150E	104	Philadelphia, PA (city)	40N 75W	86
Oklahoma (st., US)	36N 95W	86	Philippine Sea	15N 125E	104
Oklahoma City, Oklahoma (city, st. cap., US)	35N 98W	86	Philippines (country)	15N 120E	103
Oland (island)	57N 17E	95	Phnom Penh, Cambodia (city, nat. cap.)	12N 105E	103
Olympia, Washington (city, st. cap., US)	47N 123W	86	Phoenix, Arizona (city, st. cap., US)	33N 112W	86
Omaha, NE (city)	41N 96W	86	Phou Bia 9,249	24N 102E	104
Oman (country)	20N 55E	103	Piaui (st., Brazil)	7S 44W	90
Oman, Gulf of	23N 55E	104	Piaui Range	10S 45W	91
Omdurman, Sudan (city)	16N 32E	99	Pic Touside 10,712	20N 12E	100
Omsk, Russia (city)	55N 73E	103	Pierre, South Dakota (city, st. cap., US)	44N 100W	86
Onega, Lake	62N 35E	95	Pietermaritzburg, South Africa (city)	30S 30E	99
Ontario (lake, N.Am.)	45N 77W	87	Pike's Peak 14,110	36N 110W	87
Ontario (prov., Can.)	50N 90W	86	Pilcomayo, Rio (riv., S.Am.)	23S 60W	91
Oodnadatta, Aust. (city)	28S 135E	107	Pittsburgh, PA (city)	40N 80W	86
Oran, Algeria (city)	36N 1W	99	Plateau of Iran	26N 60E	104
Oregon (st., US)	46N 120W	86	Plateau of Tibet	26N 85E	104
Orinoco, Rio (riv., S.Am.)	8N 65W	91	Platte (riv., N.Am.)	41N 105W	87
Orizaba Peak 18,406	19N 97W	87	Po (riv., Europe)	45N 12E	95
Orkney Islands	60N 0	95	Point Barrow	70N 156W	87 inset
Osaka, Japan (city)	35N 135E	103	Poland (country)	54N 20E	94
Oslo, Norway (city, nat. cap.)	60N 11W	94	Poopo, Lake	16S 67W	91
Ossa, Mt. 5,305 (Tasm.)	43S 145E	108	Popocatepetl 17,887	17N 100W	87
Ottawa, Canada (city, nat. cap.)	45N 76W	86	Port Elizabeth, South Africa (city)	34S 26E	99
Otway, Cape	40S 142W	108	Port Lincoln, Aust. (city)	35S 135E	107
Ougadougou, Burkina Faso (city, nat. cap.)	12N 2W	99	Port Moresby, Papua N. G. (city, nat. cap.)	10S 147E	107
Owen Stanley Range	9S 148E	108	Port Vila, Vanatu (city, nat. cap.)	17S 169E	107
Pachuca, Hidalgo (city, st. cap. Mex.)	20N 99W	86	Port-au-Prince, Haiti (city, nat. cap.)	19N 72W	86
Pacific Ocean	20N 115W	87	Portland, OR (city)	46N 123W	86
Pakistan (country)	25N 72E	103	Porto Alegre, R. Gr. do Sul (city, st. cap., Braz.)	30S 51W	90
Palawan (island)	10N 119E	104	Porto Novo, Benin (city, nat. cap.)	7N 3E	99
Palmas, Cape	8N 8W	100	Porto Velho, Rondonia (city, st. cap., Braz.)	9S 64W	90
Palmas, Tocantins (city, st. cap., Braz.)	10S 49W	90	Portugal (country)	38N 8W	94
Pamirs	32N 70E	104	Potomac (riv., N.Am.)	35N 75W	87
Pampas	36S 73W	91	Potosi, Bolivia (city)	20S 66W	90
Panama (country)	10N 80W	86	Prague, Czech Republic (city, nat. cap.)	50N 14E	94
Panama, Gulf of	10N 80W	86	Pretoria, South Africa (city, nat. cap.)	26S 28E	99
Panama, Panama (city, nat. cap.)	9N 80W	86	Pribilof Islands	56N 170W	87 inset
Papua, Gulf of	8S 144E	108	Prince Edward Island (prov., Can.)	50N 67W	86
Papua New Guinea (country)	6S 144E	108	Pripyat Marshes	54N 24E	95
Para (st., Brazil)	4S 54W	90	Providence, Rhode Island (city, st. cap., US)	42N 71W	86
Paraguay (country)	23S 60W	90	Puebla (st., Mex.)	18N 96W	86
Paraguay, Rio (riv., S.Am.)	17S 60W	91	Puebla, Puebla (city, st. cap., Mex.)	19N 98W	86
Paraiba (st., Brazil)	6S 35W	90	Puerto Monte, Chile (city)	42S 74W	90
Paramaribo, Suriname (city, nat. cap.)	5N 55W	90	Purus, Rio (riv., S.Am.)	5S 68W	91
Parana (st., Brazil)	25S 55W	90	Putumayo, Rio (riv., S.Am.)	3S 74W	91
Parana, Entre Rios (city, st. cap., Argen.)	32S 60W	90	Pyongyang, Korea, North (city, nat. cap.)	39N 126E	103
Parana, Rio (riv., S.Am.)	20S 50W	91	Pyrenees Mountains	43N 2E	95
Paris, France (city, nat. cap.)	49N 2E	94	Qingdao, China (city)	36N 120E	103
Pasadas, Misiones (city, st. cap., Argen.)	27S 56W	90	Quebec (prov., Can.)	52N 70W	86

Name/Description	Latitude & Longitude	Page	Name/Description	Latitude & Longitude	Page
Quebec, Quebec (city, prov. cap., Can.)	47N 71W	86	Rub al Khali	20N 50E	104
Queen Charlotte Islands	50N 130W	87	Rudolph, Lake	3N 34E	100
Queen Elizabeth Islands	75N 110W	87	Russia (country)	58N 56E	94
Queensland (st., Aust.)	24S 145E	107	Ruvuma (riv., Africa)	12S 38E	100
Querataro (st., Mex.)	22N 96W	86	Ruwenzori Mountains	0 30E	100
Querataro, Querataro (city, st. cap., Mex.)	21N 100W	86	Rwanda (country)	3S 30E	99
Quintana Roo (st., Mex.)	18N 88W	86	Rybinsk, Lake	58N 38E	95
Quito, Ecuador (city, nat. cap.)	0 79W	90	S. Saskatchewan (riv., N.Am.)	50N 110W	87
Rabat, Morocco (city, nat. cap.)	34N 7W	99	Sable, Cape	45N 70W	87
Race, Cape	46N 52W	87	Sacramento (riv., N.Am.)	40N 122W	87
Rainier, Mt. 14,410	48N 120W	87	Sacramento, California (city, st. cap. US)	39 121W	
Raleigh, North Carolina (city, st. cap., US)	36N 79W	86	Sahara	18N 10E	100
Rangoon, Myanmar (Burma) (city, nat. cap.)	17N 96E	103	Sakhalin Island	50N 143E	104
Rapid City, SD (city)	44N 103W	86	Salado, Rio (riv., S.Am.)	35S 70W	91
Rawalpindi, India (city)	34N 73E	103	Salem, Oregon (city, st. cap., US)	45N 123W	86
Rawson, Chubuy (city, st. cap., Argen.)	43S 65W	90	Salt Lake City, Utah (city, st. cap., US)	41N 112W	86
Recife, Pernambuco (city, st. cap., Braz.)	8S 35W	90	Salta (st., Argentina)	25S 70W	90
Red (of the North) (riv., N.Am.)	50N 98W	87	Salta, Salta (city, st. cap., Argen.)	25S 65W	90
Red (riv., N.Am.)	42N 96W	87	Saltillo, Coahuila (city, st. cap., Mex.)	26N 101W	86
Red Sea	20N 35E	100	Salvador, Bahia (city, st. cap., Braz.)	13S 38W	90
Regina, Canada (city)	51N 104W	86	Salween (riv., Asia)	18N 98E	104
Reindeer (lake, N.Am.)	57N 100W	87	Samar (island)	12N 124E	104
Repulse Bay	22S 147E	108	Samara, Russia (city)	53N 50E	94
Resistencia, Chaco (city, st. cap., Argen.)	27S 59W	90	Samarkand, Uzbekistan (city)	40N 67E	94
Revillagigedo Island	18N 110W	87	San Antonio, TX (city)	29N 98W	86
Reykjavik, Iceland (city, nat. cap.)	64N 22W	94	San Cristobal (island)	12S 162E	108
Rhine (riv., Europe)	50N 10E	95	San Diego, CA (city)	33N 117W	86
Rhode Island (st., US)	42N 70W	86	San Francisco, CA (city)	38N 122W	86
Rhone (riv., Europe)	42N 8E	95	San Francisco, Rio (riv., S.Am.)	10S 40W	91
Richmond, Virginia (city, st. cap., US)	38N 77W	86	San Joaquin (riv., N.Am.)	37N 121W	87
Riga, Gulf of	58N 24E	95	San Jorge, Gulf of	45S 68W	91
Riga, Latvia (city, nat. cap.)	57N 24E	94	San Jose, Costa Rica (city, nat. cap.)	10N 84W	86
Rio Branco, Acre (city, st. cap., Braz.)	10S 68W	90	San Juan (st., Argentina)	30S 70W	90
Rio de Janeiro (st., Brazil)	22S 45W	90	San Juan, San Juan (city, st. cap., Argen.)	18N 66W	86
Rio de Janeiro, R. de Jan. (city, st. cap., Braz.)	23S 43W	90	San Lucas, Cape	23N 110W	87
Rio de la Plata	35S 55W	91	San Luis Potosi (st., Mex.)	22N 101W	86
Rio Gallegos, Santa Cruz (city, st. cap., Argen.)	52S 68W	90	San Luis Potosi, S. Luis P. (city, st. cap., Mex.)	22N 101W	86
Rio Grande (riv., N.Am.)	30N 100W	87	San Matias, Gulf of	43S 65W	91
Rio Grande do Norte (st., Brazil)	5S 35W	90	San Salvador, El Salvador (city, nat. cap.)	14N 89W	86
Rio Grande do Sul (st., Brazil)	30S 55W	90	Sanaa, Yemen (city)	16N 44E	99
Rio Negro (st., Argentina)	40S 70W	90	Santa Catarina (st., Brazil)	28S 50W	90
Riyadh, Saudi Arabia (city, nat. cap.)	25N 47E	94	Santa Cruz (st., Argentina)	50S 70W	90
Roanoke (riv., N.Am.)	34N 75W	87	Santa Cruz Islands	8S 168E	108
Roberts, Mt. 4,495	28S 154E	108	Santa Fe (st., Argentina)	30S 62W	90
Rockhampton, Aust. (city)	23S 150E	107	Santa Fe de Bogota, Colombia (city, nat. cap.)	5N 74W	90
Rocky Mountains	50N 108W	87	Santa Fe, New Mexico (city, st. cap., US)	35N 106W	86
Roebuck Bay	18S 125E	108	Santa Rosa, La Pampa (city, st. cap., Argen.)	37S 64W	90
Romania (country)	46N 24E	94	Santiago, Chile (city, nat. cap.)	33S 71W	90
Rome, Italy (city, nat. cap.)	42N 13E	94	Santiago del Estero (st., Argentina)	25S 65W	90
Rondonia (st., Brazil)	12S 65W	90	Santiago, Sant. del Estero (city, st. cap., Argen.)	28S 64W	90
Roosevelt, Rio (riv., S.Am.)	10S 60W	91	Santo Domingo, Dominican Rep. (city, nat. cap.)	18N 70W	86
Roper (riv., Australasia)	15S 135W	108	Santos, Brazil (city)	24S 46W	90
Roraima (st., Brazil)	2N 62W	90	Sao Luis, Maranhao (city, st. cap., Braz.)	3S 43W	90
Ros Dashen Terrara 15,158	12N 40E	100	Sao Paulo (st., Brazil)	22S 50W	90
Rosario, Santa Fe (city, st. cap., Argen.)	33S 61W	90	Sao Paulo, Sao Paulo (city, st. cap., Braz.)	24S 47W	90
Rostov, Russia (city)	47N 40E	94	Sarajevo, Bosnia and Herz. (city, nat. cap.)	43N 18E	94
Rotterdam, Netherlands (city)	52N 4E	94	Sardinia (island)	40N 10E	95
Ruapehu, Mt. 9,177	39S 176W	108	Sarmiento, Mt. 8,100	55S 72W	91

Name/Description	Latitude & Longitude	Page	Name/Description	Latitude & Longitude	Page
Saskatchewan (riv., N.Am.)	52N 108W	87	St. Elias, Mt. 18,008	61N 139W	87
Saudi Arabia (country)	25N 50E	103	St. George's Channel	53N 5W	95
Savannah (riv., N.Am.)	33N 82W	87	St. Helena (island)	16S 5W	99
Savannah, GA (city)	32N 81W	86	St. John's, Nwfndlnd (city, prov. cap., Can.)	48N 53W	86
Sayan Range	45N 90E	104	St. Louis, MO (city)	39N 90W	86
Seattle, WA (city)	48N 122W	86	St. Lawrence (island)	65N 170W	87 inset
Seine (riv., Europe)	49N 3E	95	St. Lawrence (riv., N.Am.)	50N 65W	87
Senegal (country)	15N 15W	99	St. Lawrence, Gulf of	50N 65W	87
Senegal (riv., Africa)	15N 15W	100	St. Marie, Cape	25S 45E	99
Seoul, Korea, South (city, nat. cap.)	38N 127E	103	St. Paul, Minnesota (city, st. cap., US)	45N 93W	86
Sepik (riv., Australasia)	4S 142E	108	St. Petersburg, Russia (city)	60N 30E	94
Sergipe (st., Brazil)	12S 36W	90	St. Vincente, Cape of	37N 10W	95
Sev Dvina (riv., Asia)	60N 50E	104	Stanovoy Range	55N 125E	104
Severnaya Zemlya (island)	80N 88E	104	Stavanger, Norway (city)	59N 6E	94
Shanghai, China (city)	31N 121E	103	Steep Point	25S 115E	108
Shasta, Mt. 14,162	42N 120W	87	Stockholm, Sweden (city, nat. cap.)	59N 18E	94
Shenyang, China (city)	42N 123E	103	Stuart Range	32S 135E	108
Shetland Islands	60N 5W	95	Stuttgart, Germany (city)	49N 9E	94
Shikoku (island)	34N 130E	104	Sucre, Bolivia (city)	19S 65W	90
Shiraz, Iran (city)	30N 52E	103	Sudan (country)	10N 30E	99
Sicily (island)	38N 14E	94	Sulaiman Range	28N 70E	104
Sierra Leone (country)	6N 14W	99	Sulu Islands	8N 120E	103
Sierra Madre Occidental	27N 108W	87	Sulu Sea	10N 120E	103
Sierra Madre Oriental	27N 100W	87	Sumatra (island)	0 100E	104 inset
Sierra Nevada	38N 120W	87	Sumba (island)	10S 120E	108
Sikhote Alin	45N 135E	104	Sumbawa (island)	8S 116E	108
Simpson Desert	25S 136E	108	Sunda Islands	12S 118E	108
Sinai Peninsula	28N 33E	100	Superior (lake, N.Am.)	50N 90W	87
Sinaloa (st., Mex.)	25N 110W	86	Surabaya, Java (Indonesia) (city)	7S 113E	103 inset
Singapore (city, nat. cap.)	1N 104E	103	Suriname (country)	5N 55W	90
Sitka Island	57N 125W	87	Svalbard Islands	75N 20E	104
Skagerrak, Strait of	58N 8E	95	Swan (riv., Australasia)	34S 115E	108
Skopje, Macedonia (city, nat. cap.)	42N 21E	94	Sweden (country)	62N 16E	94
Slovakia (country)	50N 20E	94	Sydney, N.S.Wales (city, st. cap., Aust.)	34S 151E	107
Slovenia (country)	47N 14E	94	Syr Darya (riv., Asia)	36N 65E	104
Snake (riv., N.Am.)	45N 110W	87	Syria (country)	37N 36E	94
Snowy Mountains	37S 148E	108	Tabasco (st., Mex.)	16N 90W	86
Sofia, Bulgaria (city, nat. cap.)	43N 23E	94	Tabriz, Iran (city)	38N 46E	103
Solimoes, Rio (riv., S.Am.)	3S 65W	91	Tahat, Mt. 9,541	23N 8E	100
Solomon Islands (country)	7S 160E	108	Taipei, Taiwan (city, nat. cap.)	25N 121E	103
Somalia (country)	5N 45E	99	Taiwan (country)	25N 122E	103
Sonora (st., Mex.)	30N 110W	86	Taiwan Strait	25N 120E	104
South Africa (country)	30S 25E	99	Tajikistan (country)	35N 75E	103
South Australia (st., Aust.)	30S 125E	107	Takla Makan	37N 90E	104
South Cape, New Guinea	8S 150E	108	Tallahassee, Florida (city, st. cap., US)	30N 84W	86
South Carolina (st., US)	33N 79W	86	Tallinn, Estonia (city, nat. cap.)	59N 25E	94
South China Sea	15N 115E	108	Tamaulipas (st., Mex.)	25N 95W	86
South Dakota (st., US)	45N 100W	86	Tampico, Mexico (city)	22N 98W	86
South Georgia (island)	55S 40W	91	Tanganyika, Lake	5S 30E	100
South Island (NZ)	45S 170E	108	Tanzania (country)	8S 35E	99
Southampton Island	68N 86W	87	Tapajos, Rio (riv., S.Am.)	5S 55W	91
Southern Alps (NZ)	45S 170E	108	Tarim Basin	37N 85E	104
Southwest Cape (NZ)	47S 167E	108	Tashkent, Uzbekistan (city, nat. cap.)	41N 69E	103
Spain (country)	38N 4W	94	Tasman Sea	38S 160E	108
Spokane, WA (city)	48N 117W	86	Tasmania (st., Aust.).	42S 145E	107
Springfield, Illinois (city, st. cap., US)	40N 90W	86	Tatar Strait	50N 142E	104
Sri Lanka (country)	8N 80E	103	Tbilisi, Georgia (city, nat. cap.)	42N 45E	94
Srinagar, India (city)	34N 75E	103	Teguicigalpa, Honduras (city, nat. cap.)	14N 87W	86

Name/Description	Latitude & Longitude	Page	Name/Description	Latitude & Longitude	Page
Tehran, Iran *(city, nat. cap.)*	36N 51E	103	Tuxtla Gutierrez, Chiapas *(city, st. cap., Mex.)*	17N 93W	86
Tel Aviv, Israel *(city)*	32N 35E	94	Tyrrhenian Sea	40N 12E	95
Tennant Creek, Aust. *(city)*	19S 134E	107	Ubangi *(riv., Africa)*	0 20E	100
Tennessee *(st., US)*	37N 88W	86	Ucayali, Rio *(riv., S.Am.)*	7S 75W	91
Tennessee *(riv., N.Am.)*	32N 88W	87	Uele *(riv., Africa)*	3N 25E	100
Tepic, Nayarit *(city, st. cap., Mex.)*	22N 105W	86	Uganda *(country)*	3N 30E	99
Teresina, Piaui *(city, st. cap., Braz.)*	5S 43W	90	Ujungpandang, Celebes (Indon.) *(city)*	5S 119E	103 inset
Texas *(st., US)*	30N 95W	86	Ukraine *(country)*	53N 32E	94
Thailand *(country)*	15N 105E	103	Ulan Bator, Mongolia *(city, nat. cap.)*	47N 107E	103
Thailand, Gulf of	10N 105E	104	Uliastay, Mongolia *(city)*	48N 97E	103
Thames *(riv., Europe)*	52N 4W	95	Ungava Peninsula	60N 72W	87
The Hague, Netherlands *(city, nat. cap.)*	52N 4E	94	United Arab Emirates *(country)*	25N 55E	103
The Round Mountain 5,300	29S 152E	108	United Kingdom *(country)*	54N 4W	94
Thimphu, Bhutan *(city, nat. cap.)*	28N 90E	103	United States *(country)*	40N 100W	86
Tianjin, China *(city)*	39N 117E	103	Uppsala, Sweden *(city)*	60N 18E	94
Tibest Massif	20N 20E	100	Ural *(riv., Asia)*	45N 55E	104
Tien Shan	40N 80E	104	Ural Mountains	50N 60E	104
Tierra del Fuego	54S 68W	91	Uruguay *(country)*	37S 67W	90
Tierra del Fuego *(st., Argentina)*	54S 68W	90	Uruguay, Rio *(riv., S.Am.)*	30S 57W	91
Tigris *(riv., Asia)*	37N 40E	95	Urumqui, China *(city)*	44N 88E	103
Timor *(island)*	7S 126E	104	Utah *(st., US)*	38N 110W	86
Timor Sea	11S 125E	107	Uzbekistan *(country)*	42N 58E	94
Tirane, Albania *(city, nat. cap.)*	41N 20E	94	Vaal *(riv., Africa)*	27S 27E	100
Titicaca, Lake	15S 70W	91	Valdivia, Chile *(city)*	40S 73W	90
Tlaxcala *(st., Mex.)*	20N 96W	86	Valencia, Spain *(city)*	39N 0	94
Tlaxcala, Tlaxcala *(city, st. cap., Mex.)*	19N 98W	86	Valencia, Venezuela *(city)*	10N 68W	90
Toamasino, Madagascar *(city)*	18S 49E	99	Valparaiso, Chile *(city)*	33S 72W	90
Tocantins *(st., Brazil)*	12S 50W	90	van Diemen, Cape	11S 130E	108
Tocantins, Rio *(riv., S.Am.)*	5S 50W	91	van Rees Mountains	4S 140E	108
Togo *(country)*	8N 1E	99	Vanatu *(country)*	15S 167E	107
Tokyo, Japan *(city, nat. cap.)*	36N 140E	103	Vancouver, Canada *(city)*	49N 123W	86
Toliara, Madagascar *(city)*	23S 44E	99	Vancouver Island	50N 130W	87
Tolima, Mt. 17,110	5N 75W	91	Vanern, Lake	60N 12E	95
Toluca, Mexico *(city, st. cap., Mex.)*	19N 100W	86	Vattern, Lake	56N 12E	95
Tombouctou, Mali *(city)*	24N 3W	99	Venezuela *(country)*	5N 65W	90
Tomsk, Russia *(city)*	56N 85E	103	Venezuela, Gulf of	12N 72W	91
Tonkin, Gulf of	20N 108E	104	Venice, Italy *(city)*	45N 12E	94
Topeka, Kansas *(city, st. cap., US)*	39N 96W	86	Vera Cruz *(st., Mex.)*	20N 97W	86
Toronto, Ontario *(city, prov. cap., Can.)*	44N 79W	86	Vera Cruz, Mexico *(city)*	19N 96W	86
Toros Mountains	37N 45E	104	Verkhoyanskiy Range	65N 130E	104
Torrens, Lake	33S 136W	108	Vermont *(st., US)*	45N 73W	86
Torres Strait	10S 142E	108	Vert, Cape	15N 17W	100
Townsville, Aust. *(city)*	19S 146E	107	Vestfjord	68N 14E	95
Transylvanian Alps	46N 20E	95	Viangchan, Laos *(city, nat. cap.)*	18N 103E	103
Trenton, New Jersey *(city, st. cap., US)*	40N 75W	86	Victoria *(riv., Australasia)*	15S 130E	108
Tricara Peak 15,584	4S 137E	108	Victoria *(st., Aust.)*	37S 145W	107
Trinidad and Tobago *(island)*	9N 60W	91	Victoria, B.C. *(city, prov. cap., Can.)*	48N 123W	86
Tripoli, Libya *(city, nat. cap.)*	33N 13E	99	Victoria, Lake	3S 35E	100
Trujillo, Peru *(city)*	8S 79W	90	Victoria, Mt. 13,238	9S 137E	108
Tucson, AZ *(city)*	32N 111W	86	Victoria Riv. Downs, Aust. *(city)*	17S 131E	107
Tucuman *(st., Argentina)*	25S 65W	90	Viedma, Rio Negro *(city, st. cap., Argen.)*	41S 63W	90
Tucuman, Tucuman *(city, st. cap., Argen.)*	27S 65W	90	Vienna, Austria *(city)*	48N 16E	94
Tunis, Tunisia *(city, nat. cap.)*	37N 10E	99	Vietnam *(country)*	10N 110E	103
Tunisia *(country)*	34N 9E	99	Villahermosa, Tabasco *(city, st. cap., Mex.)*	18N 93W	86
Turin, Italy *(city)*	45N 8E	94	Vilnius, Lithuania *(city, nat. cap.)*	55N 25E	94
Turkey *(country)*	39N 32E	94	Virginia *(st., US)*	35N 78W	86
Turkmenistan *(country)*	39N 56E	94	Viscount Melville Sound	72N 110W	87
Turku, Finland *(city)*	60N 22E	94	Vitoria, Espiritu Santo *(city, st. cap., Braz.)*	20S 40W	90

Name/Description	Latitude & Longitude	Page	Name/Description	Latitude & Longitude	Page
Vladivostock, Russia *(city)*	43N 132E	103	Wyndham, Australia *(city)*	16S 129E	107
Volga *(riv., Europe)*	46N 46E	95	Wyoming *(st., US)*	45N 110W	86
Volgograd, Russia *(city)*	54N 44E	94	Xalapa, Vera Cruz *(city, st. cap., Mex.)*	20N 97W	86
Volta *(riv., Africa)*	10N 15E	100	Xingu, Rio *(riv., S.Am.)*	5S 54W	91
Volta, Lake	8N 2W	100	Yablonovyy Range	50N 100E	104
Vosges Mountains	48N 7E	95	Yakutsk, Russia *(city)*	62N 130E	103
Wabash *(riv., N.Am.)*	43N 90W	87	Yamoussoukio, Cote d'Ivoire *(city)*	7N 4W	99
Walvis Bay, Namibia *(city)*	23S 14E	99	Yangtze (Chang Jiang) *(riv., Asia)*	30N 108W	104
Warsaw, Poland *(city, nat. cap.)*	52N 21E	94	Yaounde, Cameroon *(city, nat. cap.)*	4N 12E	99
Washington *(st., US)*	48N 122W	86	Yekaterinburg, Russia *(city)*	57N 61E	94
Washington, D.C., United St.s *(city, nat. cap.)*	39N 77W	86	Yellowknife, N.W.T. *(city, prov. cap., Can.)*	62N 115W	86
Wellington Island	48S 74W	91	Yellowstone *(riv., N.Am.)*	46N 110W	87
Wellington, New Zealand *(city, nat. cap.)*	41S 175E	107	Yemen *(country)*	15N 50E	103
Weser *(riv., Europe)*	54N 8E	95	Yenisey *(riv., Asia)*	68N 85E	104
West Cape Howe	36S 115E	108	Yerevan, Armenia *(city, nat. cap.)*	40N 44E	94
West Indies	18N 75W	87	Yokohama, Japan *(city)*	36N 140E	103
West Siberian Lowland	60N 80E	104	York, Cape	75N 65W	87
West Virginia *(st., US)*	38N 80W	86	Yucatan *(st., Mex.)*	20N 88W	86
Western Australia *(st., Aust.)*	25S 122W	107	Yucatan Channel	22N 88W	87
Western Ghats	15N 72E	104	Yucatan Peninsula	20N 88W	87
Western Sahara *(country)*	25N 13W	99	Yugoslavia *(country)*	44N 20E	94
White Nile *(riv., Africa)*	13N 30E	100	Yukon *(riv., N.Am.)*	63N 150W	87 inset
White Sea	64N 36E	95	Zacatecas *(st., Mex.)*	23N 103W	86
Whitney, Mt. 14,494	33N 118W	87	Zacatecas, Zacatecas *(city, st. cap., Mex.)*	23N 103W	86
Wichita, KS *(city)*	38N 97W	86	Zagreb, Croatia *(city, nat. cap.)*	46N 16E	94
Wilhelm, Mt. 14,793	4S 145E	108	Zagros Mountains	27N 52E	104
Windhoek, Namibia *(city, nat. cap.)*	22S 17E	99	Zambezi *(riv., Africa)*	18S 30E	100
Winnipeg *(lake, N.Am.)*	50N 100W	87	Zambia *(country)*	15S 25E	99
Winnipeg, Manitoba *(city, prov. cap., Can.)*	53N 98W	86	Zanzibar *(island)*	5S 39E	100
Wisconsin *(st., US)*	50N 90W	86	Zemlya Frantsa Josifa *(island)*	80N 40E	104
Wollongong, Aust. *(city)*	34S 151E	107	Ziel, Mt. 4,955	23S 134E	108
Woodroffe, Mt. 4,724	26S 133W	108	Zimbabwe *(country)*	20S 30E	99
Woomera, Aust. *(city)*	32S 137E	107			
Wrangell *(island)*	72N 180E	104			
Wuhan, China *(city)*	30N 114E	103			

Sources

After the storm. (1991, August). *National Geographic,* 180.

Alaska's big spill. (1990, January). *National Geographic,* 177.

Amazonia [map]. (1994). *National Geographic,* 186.

An atmosphere of uncertainty. (1987, April). *National Geographic,* 171.

Crabb, C. (1993, January). Soiling the planet. *Discover, 14*(1), 74–75.

DeBlij, H. J., & Muller, P. (1997). *Geography: Realms, regions and concepts* (8th ed., revised). New York: John Wiley & Sons.

Department of Geography, Pennsylvania State University. (1996). Unpublished computer model output. State College, PA: Pennsylvania State University.

Domke, K. (1988). *War and the changing global system.* New Haven, CT: Yale University Press.

Eastern Europe's dark dawn. (1991, June). *National Geographic,* 179.

Economic consequences of the accident at Chernobyl nuclear plant. (1987). PlanEcon Reports, 3.

Environmental Protection Agency. (1996). Unpublished data [online]. Available: http://www.epa.gov.

Fellman, J., Getis, A., & Getis, J. (1995). *Human geography: Landscapes of human activities* (4th ed.). Dubuque, IA: Wm. C. Brown Publishers.

Fuller, Harold (ed.). (1971). *World Patterns: The Aldine College Atlas.* Chicago: Aldine Publishing Co.

Johnson, D. (1977). *Population, society, and desertification.* New York: United Nations Conference on Desertification, United Nations Environment Programme.

Köppen, W., & Geiger, R. (1954). *Klima der erde* [Climate of the earth]. Darmstadt, Germany: Justus Perthes.

Kuchler, A. W. (1949). Natural vegetation. *Annals of the Association of American Geographers,* 39.

Lindeman, M. (1990). *The United States and the Soviet Union: Choices for the 21st century.* Guilford, CT: Dushkin Publishing Group.

Mather, J. R. (1974). *Climatology: Fundamentals and applications.* New York: McGraw-Hill.

Miller, G. T. (1992). *Living in the environment* (7th ed.). Belmont, CA: Wadsworth.

Murphy, R. E. (1968). Landforms of the world [Map supplement No. 91]. *Annals of the Association of American Geographers, 58*(1), 198–200.

National Aeronautics and Space Administration. (1994–1996). Unpublished data and images [online]. Available: http://www.nasa.gov.

National Geographic Society. (1995). *Atlas of the world,* revised 6th edition. Washington, D.C.: National Geographic Society.

National Oceanic and Atmospheric Administration. (1996). Unpublished data [online]. Available: http://www.noaa.gov.

The Oglalla Aquifer. (1993, March). *National Geographic,* 183.

Population Reference Bureau. (1994). *1994 world population data sheet.* New York: Population Reference Bureau.

Rand McNally. (1996). *Goode's world atlas* (19th ed.). Chicago: Rand McNally and Co.

Rand McNally answer atlas. (1996). Chicago: Rand McNally and Co.

Rondonia: Brazil's imperiled rainforest. (1988, December). *National Geographic,* 174.

Rourke, J. T. (1999). *International politics on the world stage* (7th ed). Guilford, CT: Dushkin/McGraw-Hill.

Shelley, F., & Clarke, A. (1994). *Human and cultural geography: A global perspective,* Dubuque, IA: Wm. C. Brown Publishers.

Smith, D. (1997). *The state of war and peace atlas,* 3rd edition. New York: Penguin Books.

Soiling the planet. (1993, January). *Discover,* 14.

Spector, L. S., & Smith, J. R. (1990). *Nuclear ambitions: The spread of nuclear weapons.* Boulder, CO: Westview Press.

This fragile earth [map]. (1988, December). *National Geographic,* 174.

Thornthwaite, C. W., & Mather, J. R. (1955). *The water balance* [Publications in Climatology No. 8]. Centerton, NJ: Drexel Institute of Technology, Laboratory of Climatology.

Times atlas of world history. (1978). Maplewood, NJ: Hammond.

United Nations Food and Agriculture Organization (FAO). (1995). *Forest resources assessment 1990: Global synthesis* [FAO Forestry Paper No. 124]. Rome: FAO.

United Nations Population Fund. (1992). *The state of the world's population.* New York: United Nations Population Fund.

United Nations Population Reference Bureau. (1990). *World development report.* New York: Oxford University Press.

U.S. Census Bureau. (1994). *World population profile.* Washington, D.C.: U.S. Government Printing Office.

U.S. Central Intelligence Agency. (1995). *World factbook 1995.* Available: http://www.odci.gov/cia/publications/95fact/index html.

U.S. Central Intelligence Agency. (1996). Unpublished data [online]. Available: http://www.odci.gov/cia/publications.

U.S. Central Intelligence Agency. (1996). *World factbook 1996–97.* Washington, D.C.: Brasseys.

U.S. Committee for Refugees. (1996). *World refugee survey.* Washington, D.C.: U.S. Government Printing Office.

U.S. Department of Energy. (1996). *U.S.–Canada memorandum of intent on transboundary air pollution.* Washington, D.C.: U.S. Government Printing Office.

U.S. Department of State. (1997). *Statesman's year-book, 1997–98.* Washington, D.C.: U.S. Government Printing Office.

U.S. Forest Service. (1989). *Ecoregions of the continents.* Washington, D.C.: U.S. Government Printing Office.

U.S. Soil Conservation Service [now the U.S. Natural Resources Conservation Service]. (1996). *World soils.* Washington, D.C.: U.S. Soil Conversation Service.

The World Almanac and Book of Facts 1998. (1998). Mahwah, NJ: World Almanac Books.

The World Bank. (1995). *World development report 1995.* Geneva: World Bank.

The World Bank. (1998). *1998 World development indicators.* Washington, D.C.: World Bank.

World Conservation Monitoring Centre. (1996). Unpublished data, Cambridge, England: World Conservation Monitoring Centre.

World Health Organization. (1997). *World health statistics annual.* Geneva: World Health Organization.

World Resources Institute. *World resources 1998–99: A guide to the global environment.* Oxford University Press: New York, 1998.

Worldwatch Institute. (1987). *Reassessing nuclear power: The fallout from Chernobyl* [Worldwatch paper 75]. New York: Worldwatch Institute.

Wright, John W. (ed.). (1998). *The New York Times 1998 Almanac.* New York: Penguin Reference Books.